# The Black Earth

# International Year of Planet Earth

**Series Editors:**

Eduardo F.J. de Mulder
Executive Director International Secretariat
International Year of Planet Earth

Edward Derbyshire
Goodwill Ambassador
International Year of Planet Earth

The book series is dedicated to the United Nations International Year of Planet Earth. The aim of the Year is to raise worldwide public and political awareness of the vast (but often under-used) potential of Earth sciences for improving the quality of life and safeguarding the planet. Geoscientific knowledge can save lives and protect property if threatened by natural disasters. Such knowledge is also needed to sustainably satisfy the growing need for Earth's resources by more people. Earth scientists are ready to contribute to a safer, healthier and more prosperous society. IYPE aims to develop a new generation of such experts to find new resources and to develop land more sustainably.

For further volumes:
http://www.springer.com/series/8096

Igori Arcadie Krupenikov ·
Boris P. Boincean · David Dent

# The Black Earth

Ecological Principles for
Sustainable Agriculture
on Chernozem Soils

Igori Arcadie Krupenikov
Stefan cel Mare Str. 130
2012 Chisinau
Republic of Moldova

Boris P. Boincean
Calea Iesilor 28
3101 Balti
Republic of Moldova
borisboincean@gmail.com

David Dent
Chestnut Tree Farmhouse
Forncett End
NR16 1HT Norfolk
United Kingdom
dentsinengland@hotmail.com

ISBN 978-94-007-0158-8          e-ISBN 978-94-007-0159-5
DOI 10.1007/978-94-007-0159-5
Springer Dordrecht Heidelberg London New York

Library of Congress Control Number: 2011922259

© Springer Science+Business Media B.V. 2011
No part of this work may be reproduced, stored in a retrieval system, or transmitted in any form or by any means, electronic, mechanical, photocopying, microfilming, recording or otherwise, without written permission from the Publisher, with the exception of any material supplied specifically for the purpose of being entered and executed on a computer system, for exclusive use by the purchaser of the work.

Printed on acid-free paper

Springer is part of Springer Science+Business Media (www.springer.com)

*Dedicated to V.V. Dokuchaev, P.A. Kostichev and V.R. Viliams, founders of the sciences of soils and agriculture*

# Foreword

The International Year of Planet Earth (IYPE) was established as a means of raising worldwide public and political awareness of the vast, though frequently under-used, potential the earth sciences possess for improving the quality of life of the peoples of the world and safeguarding Earth's rich and diverse environments.

The International Year project was jointly initiated in 2000 by the International Union of Geological Sciences (IUGS) and the Earth Science Division of the United Nations Educational, Scientific and Cultural Organisation (UNESCO). IUGS, which is a non-governmental organization, and UNESCO, an inter-governmental organization, already shared a long record of productive cooperation in the natural sciences and their application to societal problems, including the International Geoscience Programme (IGCP) now in its fourth decade.

With its main goals of raising public awareness of and enhancing research in the Earth sciences on a global scale in both the developed and less-developed countries of the world, two operational programmes were demanded. In 2002 and 2003, the series editors together with Dr. Ted Nield and Dr. Henk Schalke (all four being core members of the Management Team at that time) drew up outlines of a science and an outreach programme. In 2005, following the UN proclamation of 2008 as the United Nations International Year of Planet Earth, the "year" grew into a triennium (2007–2009).

The outreach programme, targeting all levels of human society from decision makers to the general public, achieved considerable success in the hands of member states representing over 80% of the global population. The science programme concentrated on bringing together like-minded scientists from around the world to advance collaborative science in a number of areas of global concern. A strong emphasis on enhancing the role of the Earth sciences in building a healthier, safer and wealthier society was adopted – as declared in the Year's logo strap-line "Earth Sciences *for* Society".

The organizational approach adopted by the science programme involved recognition of 10 global themes that embrace a broad range of problems of widespread national and international concern, as follows:

- Human health: this theme involves improving understanding of the processes by which geological materials affect human health as a means identifying and reducing a range of pathological effects.
- Climate: particularly emphasizes improved detail and understanding of the nonhuman factor in climate change.

- Groundwater: considers the occurrence, quantity and quality of this vital resource for all living things against a background that includes potential political tension between competing neighbour nations.
- Ocean: aims to improve understanding of the processes and environment of the ocean floors with relevance to the history of planet Earth and the potential for improved understanding of life and resources.
- Soils: this thin "skin" on Earth's surface is the vital source of nutrients that sustain life on the world's landmasses, but this living skin is vulnerable to degradation if not used wisely. This theme emphasizes greater use of soil science information in the selection, use and ensuring sustainability of agricultural soils so as to enhance production and diminish soil loss.
- Deep Earth: in view of the fundamental importance of deep the Earth in supplying basic needs, including mitigating the impact of certain natural hazards and controlling environmental degradation, this theme concentrates on developing scientific models that assist in the reconstruction of past processes and the forecasting of future processes that take place in the solid Earth.
- Megacities: this theme is concerned with means of building safer structures and expanding urban areas, including utilization of subsurface space.
- Geohazards: aims to reduce the risks posed to human communities by both natural and human-induced hazards using current knowledge and new information derived from research.
- Resources: involves advancing our knowledge of Earth's natural resources and their sustainable extraction.
- Earth and Life: it is over $2\frac{1}{2}$ billion years since the first effects of life began to affect Earth's atmosphere, oceans and landmasses. Earth's biological "cloak", known as the biosphere, makes our planet unique but it needs to be better known and protected. This theme aims to advance understanding of the dynamic processes of the biosphere and to use that understanding to help keep this global life-support system in good health for the benefit of all living things.

The first task of the leading Earth scientists appointed as theme leaders was the production of a set of theme brochures. Some 3500 of these were published, initially in English only but later translated into Portuguese, Chinese, Hungarian, Vietnamese, Italian, Spanish, Turkish, Lithuanian, Polish, Arabic, Japanese and Greek. Most of these were published in hard copy and all are listed on the IYPE web site.

It is fitting that, as the International Year's triennium terminates at the end of 2009, the more than 100 scientists who participated in the 10 science themes should bring together the results of their wide ranging international deliberations in a series of state-of-the-art volumes that will stand as a legacy of the International Year of Planet Earth. The book series was a direct result of interaction between the International Year and the Springer Verlag Company, a partnership which was formalized in 2008 during the acme of the triennium.

This IYPE-Springer book series contains the latest thinking on the chosen themes by a large number of earth science professionals from around the world. The books are written at the advanced level demanded by a potential readership consisting of Earth science professionals and students. Thus, the series is a legacy of the science programme, but it is also a counterweight to the earth science information in several media formats already delivered by the numerous national committees of the

International Year in their pursuit of worldwide popularization under the outreach programme.

The discerning reader will recognize that the books in this series provide not only a comprehensive account of the individual themes but also share much common ground that makes the series greater than the sum of the individual volumes. It is to be hoped that the scientific perspective thus provided will enhance the reader's appreciation of the nature and scale of earth science as well as the guidance it can offer to governments, decision makers and others seeking solutions to national and global problems, thereby improving everyday life for present and future residents of planet Earth.

Eduardo F.J. de Mulder
Executive Director International Secretariat
International Year of Planet Earth

Edward Derbyshire
Goodwill Ambassador
International Year of Planet Earth

# Series Preface

This book series is one of the many important results of the International Year of Planet Earth (IYPE), a joint initiative of UNESCO and the International Union of Geological Sciences (IUGS), launched with the aim of ensuring greater and more effective use by society of the knowledge and skills provided by the earth sciences.

It was originally intended that the IYPE would run from the beginning of 2007 until the end of 2009, with the core year of the triennium (2008) being proclaimed as a UN Year by the United Nations General Assembly. During all 3 years, a series of activities included in the IYPE's science and outreach programmes had a strong mobilizing effect around the globe, not only among earth scientists but also within the general public and, especially, among children and young people.

The outreach programme has served to enhance cooperation among earth scientists, administrators, politicians and civil society and to generate public awareness of the wide ranging importance of the geosciences for human life and prosperity. It has also helped to develop a better understanding of Planet Earth and the importance of this knowledge in building of a safer, healthier, and wealthier society.

The scientific programme, focused upon 10 themes of relevance to society, has successfully raised geoscientists' awareness of the need to develop further the international coordination of their activities. The programme has also led to some important updating of the main challenges the geosciences are, and will be confronting within an agenda closely focused on societal benefit.

An important outcome of the work of the IYPE's scientific themes includes this thematic book as one of the volumes making up the IYPE-Springer Series, which was designed to provide an important element of the legacy of the International Year of Planet Earth. Many prestigious scientists, drawn from different disciplines and with a wide range of nationalities, are warmly thanked for their contributions to a series of books that epitomize the most advanced, up-to-date and useful information on evolution and life, water resources, soils, changing climate, deep earth, oceans, non-renewable resources, earth and health, natural hazards, and megacities.

This legacy opens a bridge to the future. It is published in the hope that the core message and the concerted actions of the International Year of Planet Earth throughout the triennium will continue and, ultimately, go some way toward helping to establish an improved equilibrium between human society and its home planet. As stated by the Director General of UNESCO, Koichiro Matsuura, "Our knowledge of the Earth system is our insurance policy for the future of our planet". This book series is an important step in that direction.

R. Missotten
Chief, Global Earth Observation Section
UNESCO

Alberto C. Riccardi
President
IUGS

# Preface

The leading authors are a pedologist and an agronomist. For a long time, they have been concerned about the state of agriculture and the degradation of our common wealth: the *chernozem* – the rich black earth that has supported the people of eastern Europe for generations and which, globally, still represents the biggest reserve of future arable land. They have joined together to seek solutions to the issues of land degradation and agriculture's dependence of on energy and other non-renewable resources that are increasingly scarce all over the world.

Their vantage points are the Rothamsted of the steppe: Selectia Experimental Station established 65 years ago at Balti and the Nicolai Dimo Institute in Chisinau (Kishinev) – in the heart of the province where modern soil science began with the work of V.V. Dokuchaev in the 1880s, which inspired the classical Russian school of agronomy under Viliams and was the homeland of the hardy red wheat, and the hardy farmers that grew it, that transformed the Prairies of North America. Heirs to these great traditions, the authors address very present global issues: the sustainability of agriculture and the neglected but critical role of soils as a carbon source and sink; as much as one-third of the excess greenhouse gases in the atmosphere has come from the soils, through land use change over the last century. The tract of chernozem with its extraordinary thickness and natural humus content has contributed more than most but, as the best soil in the world, it offers the best opportunity to put that carbon back in the soil – where it is needed!

Soil degradation is a global issue on a par with climatic change and loss of biodiversity. In eastern Europe, it has become more acute in the transition to a market economy. Existing farming systems match neither the processes of soil formation nor the functional principles of agro-ecosystems. This can only lead to economic and natural disaster but we optimistically consider that the situation can be turned around – if urgent and active measures are taken to restore soil fertility. The grounds for optimism are that the chernozem, though much degraded, has kept its innate capacity to build up fertility – the inherent capacity for self-aggregation and rich mineralogical inheritance from their parent material, the thick soil profile and unique quality of humus. We present the experimental data that underpin this conviction, which include a wealth of data from long-term field experiments at the Selectia Experimental Station.

Future intensification of agriculture will depend on what we are calling *ecological agriculture*: not a nostalgic return to organic farming but a new paradigm built upon the scientific principles of modern ecology. It draws upon all the experience of what, before recent addiction to agrochemicals and raw mechanical power, was considered good husbandry but it does not eschew the advances of modern research and development, which are surely needed. It supposes a more intensive turnover of nutrients and renewable sources of energy on every farm, built upon crop rotation with optimal tillage, fertilization, systems of weed and pest control, and mechanization; and with the full integration of perennial legumes and livestock.

The central place in this sustainable farming system belongs to soils, in which processes of synthesis and decomposition of soil organic matter drive soil fertility. In a very real sense, soils behave like living organisms and they fulfil many crucial roles in the biosphere: in the water cycle, nutrient cycle and the carbon cycle that regulates global climate, and which have been dangerously perturbed. It is impossible to maintain soils as the Earth's living skin without an holistic approach and the book describes the various characteristics of soils that arise from the continual interaction between their living and non-living facets. It also requires radically different policies. We know that the required reforms need the support of governments – and politics is the art of the possible – but solutions to the problems of the land must respect ecological laws that are no respecters of the economic and political imperatives of the day; there can be no compromise with ecology.

There is technical knowledge enough to recoup the health of our soils and agriculture and, essentially, this book is technical. But it is written for everyone involved with agriculture: agronomists, agricultural engineers, farmers and farm managers, executives in central and local government and earth scientists in allied fields. After all, this is everybody's business. The authors combine pedology and agronomy, drawing upon both new and old, but not obsolete, literature and resources – including their own personal knowledge. David Dent has joined them as contributing editor of this English edition with the aim of making this knowledge, and the literature of the Russian school, accessible to specialists and laymen worldwide. Its focus is Moldova but the issues are the same in the neighbouring countries that share the same soils and environment and much of the same history – Hungary, Romania, the Ukraine, south Russia and the North Caucasus, extending eastwards into central Asia. The principles apply equally to the black earths of the prairies and the pampas.

We thank everyone who has helped to publish this book. Special thanks are due to Mr Constantin Mihailescu, formerly Minister of Ecology of the Republic of Moldova; the Ecological Agency *North* of the Ministry of Ecology, Director I. Gavdiuc and the series editors Edward Derbyshire and Ed de Mulder. We are indebted to Academician Ursu for the illustrations of soils and vegetation in Moldova, Dr Freddy Nachtergaele for the global map of black earths, Mrs Y. Karpes-Liem for transcribing the English manuscript and to Larisa Mrug, Lilian and Cristina Puscas for technical assistance with the pictures.

| | |
|---|---|
| Chisinau | Igori Arcadie Krupenikov |
| Balti | Boris P. Boincean |
| Norwich | David Dent |
| April 2010 | |

# Contents

**Part I     The State of the Land and Its Management** . . . . . . . . . . . 1

**Introduction** . . . . . . . . . . . . . . . . . . . . . . . . . . . . . . . . . 3
In Essence . . . . . . . . . . . . . . . . . . . . . . . . . . . . . . . . . . . 3
A Short History . . . . . . . . . . . . . . . . . . . . . . . . . . . . . . . . 5
References . . . . . . . . . . . . . . . . . . . . . . . . . . . . . . . . . . . 7

**Part II    The Anatomy of the Black Earth** . . . . . . . . . . . . . . . . . 9

**The Soil Cover** . . . . . . . . . . . . . . . . . . . . . . . . . . . . . . . 11
The Pattern of Soil Cover . . . . . . . . . . . . . . . . . . . . . . . . . . 11
Sui Generis . . . . . . . . . . . . . . . . . . . . . . . . . . . . . . . . . . 15
Soil Sustainability . . . . . . . . . . . . . . . . . . . . . . . . . . . . . . 17
References . . . . . . . . . . . . . . . . . . . . . . . . . . . . . . . . . . . 18

**Soil Texture and Structure** . . . . . . . . . . . . . . . . . . . . . . . . 19
Soil Texture . . . . . . . . . . . . . . . . . . . . . . . . . . . . . . . . . . 19
Soil Structure . . . . . . . . . . . . . . . . . . . . . . . . . . . . . . . . . 20
References . . . . . . . . . . . . . . . . . . . . . . . . . . . . . . . . . . . 25

**Soil Mineralogy and Elemental Composition** . . . . . . . . . . . . . . 27
Mineralogy . . . . . . . . . . . . . . . . . . . . . . . . . . . . . . . . . . 27
Total Chemical Analysis . . . . . . . . . . . . . . . . . . . . . . . . . . . 29
References . . . . . . . . . . . . . . . . . . . . . . . . . . . . . . . . . . . 32

**Adsorption Capacity and Reaction** . . . . . . . . . . . . . . . . . . . . 33
Adsorption Capacity . . . . . . . . . . . . . . . . . . . . . . . . . . . . . 33
Reaction . . . . . . . . . . . . . . . . . . . . . . . . . . . . . . . . . . . . 34
References . . . . . . . . . . . . . . . . . . . . . . . . . . . . . . . . . . . 35

**Soluble Salts and Soil Solution** . . . . . . . . . . . . . . . . . . . . . . 37
Soluble Salts . . . . . . . . . . . . . . . . . . . . . . . . . . . . . . . . . 37
Soil Solution . . . . . . . . . . . . . . . . . . . . . . . . . . . . . . . . . 38
References . . . . . . . . . . . . . . . . . . . . . . . . . . . . . . . . . . . 38

**Humus – Guardian of Fertility and Global Carbon Sink** . . . . . . . . 39
Its Significance for Farming . . . . . . . . . . . . . . . . . . . . . . . . 39
The Global Carbon Sink: Soils and Climatic Change . . . . . . . . . . . 41

Humus Stocks and Trends . . . . . . . . . . . . . . . . . . . . . . . . . . 42
    Losses of Humus Under the Plough . . . . . . . . . . . . . . . . . 43
The Quality of Humus . . . . . . . . . . . . . . . . . . . . . . . . . . . 45
References . . . . . . . . . . . . . . . . . . . . . . . . . . . . . . . . . 49

## The Nitrogen Riddle . . . . . . . . . . . . . . . . . . . . . . . . . . . . 51
Nitrogen in Soils and Agriculture . . . . . . . . . . . . . . . . . . . . . 51
Nitrates in the Environment . . . . . . . . . . . . . . . . . . . . . . . 55
References . . . . . . . . . . . . . . . . . . . . . . . . . . . . . . . . . 56

## Phosphorus and Sulphur Budgets . . . . . . . . . . . . . . . . . . . . 57
Phosphorus . . . . . . . . . . . . . . . . . . . . . . . . . . . . . . . . 57
Sulphur . . . . . . . . . . . . . . . . . . . . . . . . . . . . . . . . . . 59
References . . . . . . . . . . . . . . . . . . . . . . . . . . . . . . . . . 61

## Life in the Soil . . . . . . . . . . . . . . . . . . . . . . . . . . . . . . . 63
Soil Biodiversity . . . . . . . . . . . . . . . . . . . . . . . . . . . . . . 63
References . . . . . . . . . . . . . . . . . . . . . . . . . . . . . . . . . 66

## Soil Structure, Soil Water and Drought . . . . . . . . . . . . . . . . . 69
Biological Control on Soil Physical Attributes . . . . . . . . . . . . . . 69
Stable and Unstable Attributes . . . . . . . . . . . . . . . . . . . . . . 70
Available Water Capacity . . . . . . . . . . . . . . . . . . . . . . . . . 72
Drought . . . . . . . . . . . . . . . . . . . . . . . . . . . . . . . . . . 73
References . . . . . . . . . . . . . . . . . . . . . . . . . . . . . . . . . 75

## The Chernozem Family . . . . . . . . . . . . . . . . . . . . . . . . . . 77
Family Relationships . . . . . . . . . . . . . . . . . . . . . . . . . . . 77
*Calcareous Chernozem* . . . . . . . . . . . . . . . . . . . . . . . . . . 79
*Common Chernozem* . . . . . . . . . . . . . . . . . . . . . . . . . . 81
*Xerophyte-Wooded Chernozem* . . . . . . . . . . . . . . . . . . . . . 81
*Typical Chernozem* . . . . . . . . . . . . . . . . . . . . . . . . . . . 83
*Leached Chernozem* . . . . . . . . . . . . . . . . . . . . . . . . . . . 85
*Podzolized Chernozem* . . . . . . . . . . . . . . . . . . . . . . . . . 86
Land Evaluation and Land Degradation . . . . . . . . . . . . . . . . . 87
References . . . . . . . . . . . . . . . . . . . . . . . . . . . . . . . . . 87

## Accommodating Soil Diversity . . . . . . . . . . . . . . . . . . . . . . 89
Soil Variability and Precision Farming . . . . . . . . . . . . . . . . . . 89
References . . . . . . . . . . . . . . . . . . . . . . . . . . . . . . . . . 92

## Society's Perspective . . . . . . . . . . . . . . . . . . . . . . . . . . . 93
Soils and Society . . . . . . . . . . . . . . . . . . . . . . . . . . . . . 93
Agricultural Priorities . . . . . . . . . . . . . . . . . . . . . . . . . . . 94
References . . . . . . . . . . . . . . . . . . . . . . . . . . . . . . . . . 95

## Part III    Principles of Ecological Agriculture . . . . . . . . . . . . . 97

## Biological Cycles . . . . . . . . . . . . . . . . . . . . . . . . . . . . . . 99
Biological Cycling of Soil Organic Matter . . . . . . . . . . . . . . . . 99
Energy Balance of Arable Systems . . . . . . . . . . . . . . . . . . . . 100
Nutrient Balance of Arable Systems . . . . . . . . . . . . . . . . . . . 100
References . . . . . . . . . . . . . . . . . . . . . . . . . . . . . . . . . 103

## Contents

**Soil Health and Soil Quality** ... 105
The Imperative of Change ... 105
An Ecological Alternative ... 106
References ... 107

**Farming and Soil Health** ... 109
Natural Ecosystems and Agroecosystems ... 109
Principles of Sustainability ... 110
    Land Use Planning ... 110
    Crop Rotation ... 110
    Compost and Manure ... 116
    Minimum Tillage ... 116
Issues of Productivity ... 117
    Breaking the Law ... 117
Issues of Resilience ... 118
References ... 119

**Experimental Confirmation of the Efficiency of Different Farming Systems** ... 121
The Fertility of Chernozem ... 121
The Mathematical Equations ... 122
Changes in Soil Organic Matter Stocks and Crop Yields Under Different Farming Systems ... 126
References ... 128

**The Past, Present and Future of the Chernozem** ... 131
Curriculum Vitae ... 131
Future Scenarios ... 136
Conclusion ... 137
References ... 138

**Author Index** ... 139

**Subject Index** ... 141

# List of Figures

## Chapter 2

| | | |
|---|---|---|
| Figure 1 | Global distribution of chernozem and related soils . . . . . . . . . | 12 |
| Figure 2 | *Typical chernozem* at the V.V. Alekhin Central-Chernozem Biosphere State Reserve near Kursk, Russia . . . . . . . . . . . . | 12 |
| Figure 3 | *Kastanozem* in the short-grass steppe of southern Ukraine . . . . . | 13 |
| Figure 4 | *Dark grey forest soil* in the Russian forest-steppe zone . . . . . . . | 13 |
| Figure 5 | *Gley soil* in a broad depression within the short-grass steppe of southern Russia . . . . . . . . . . . . . . . . . . . . . . | 13 |
| Figure 6 | Levels of soil organization . . . . . . . . . . . . . . . . . . . . . | 16 |
| Figure 7 | The soil–crops–atmosphere system . . . . . . . . . . . . . . . . . | 17 |

## Chapter 7

| | | |
|---|---|---|
| Figure 1 | The global carbon cycle . . . . . . . . . . . . . . . . . . . . . . . | 41 |
| Figure 2 | Composition of humus in Moldavian soils . . . . . . . . . . . . . | 48 |
| Figure 3 | Soil organic carbon and carbonate carbon profiles of Moldavian soils . . . . . . . . . . . . . . . . . . . . . . . . . . . . | 49 |

## Chapter 8

| | | |
|---|---|---|
| Figure 1 | Localization of nitrates in the profiles of Moldavian soils . . . . . | 53 |
| Figure 2 | Consequences of excess nitrogen . . . . . . . . . . . . . . . . . | 55 |

## Chapter 9

| | | |
|---|---|---|
| Figure 1 | Phosphate fractions and profile distribution in chernozem, as a percentage of the total phosphate content . . . . . . . . . . . | 59 |
| Figure 2 | The sulphur cycle . . . . . . . . . . . . . . . . . . . . . . . . . . | 60 |

## Chapter 10

Figure 1    Crotovinas in chernozem . . . . . . . . . . . . . . . . . . . . . . .    64

## Chapter 12

Figure 1    *Calcareous chernozem* from the southern plain of
            Moldova: (**a**) soil profile; (**b**) steppe with *Stipa* spp.;
            (**c**) farmland . . . . . . . . . . . . . . . . . . . . . . . . . . . . . . . .    78
Figure 2    Eroded *Ordinary chernozem*, southern plain of
            Moldova: (**a**) soil profile; (**b**) association of grasses;
            (**c**) arable on sloping land . . . . . . . . . . . . . . . . . . . . .    79
Figure 3    *Typical chernozem*, low-humic phase, on the southern
            plain of Moldova: (**a**) soil profile; (**b**) natural steppe
            vegetation with *Stipa* spp . . . . . . . . . . . . . . . . . . . . . .    79
Figure 4    *Typical chernozem*, Balti steppe, Moldova: (**a**) soil profile;
            (**b**) arable; (**c**) natural steppe vegetation . . . . . . . . . . . . . . .    80
Figure 5    *Leached chernozem*, Codri, central Moldova: (**a**) soil
            profile under cultivation; (**b**) farmland . . . . . . . . . . . . . . .    80
Figure 6    *Leached chernozem* in the forest-steppe zone of northern
            Moldova: (**a**) Virgin *Leached chernozem* (**b**) *natural
            steppe vegetation* . . . . . . . . . . . . . . . . . . . . . . . . . . . .    81
Figure 7    *Grey-wooded soil* in the forest-steppe zone of northern
            Moldova: (**a**) soil profile; (**b**) Oak woodland; (**c**) farmland . . . . .    82
Figure 8    *Calcareous chernozem,* micromorphology: (**a**) direct
            light; (**b**) crossed nicols . . . . . . . . . . . . . . . . . . . . . . .    83
Figure 9    *Common chernozem*, micromorphology: (**a**) direct light;
            (**b**) crossed nicols . . . . . . . . . . . . . . . . . . . . . . . . . . .    84
Figure 10   *Typical chernozem,* micromorphology: (**a**) direct light;
            (**b**) crossed nicols . . . . . . . . . . . . . . . . . . . . . . . . . . .    85
Figure 11   *Leached chernozem*, micromorphology: (**a**) direct light;
            (**b**) crossed nicols . . . . . . . . . . . . . . . . . . . . . . . . . . .    86

## Chapter 13

Figure 1    Detailed soil map of Causheni STS. **I** (a) Layout of
            experimental fields (b) thickness of A+B horizons;
            **II** Humus content, %; **III** CaCO$_3$, %; **IV** P$_2$O$_5$, mg/100 g;
            **V** K$_2$O mg/100 g . . . . . . . . . . . . . . . . . . . . . . . . . . .    90
Figure 2    Detailed soil map of Ribnita STS. **I** (**a**) Layout of
            experimental fields (**b**) soil map; **II** (**c**) thickness of A+B
            horizons, (**d**) depth to effervescence with HCl; **III** humus
            content, %; **IV** P$_2$O$_5$, mg/100 g; **V** K$_2$O mg/100 g . . . . . . . . .    91

## Chapter 17

| | | |
|---|---|---|
| Figure 1 | "The greatest treasure of Moldavia is the vineyards." *Charles de Jopecourt 1617* | 110 |
| Figure 2 | Preparation for sowing winter wheat in the Selectia long-term field experiment with different crop rotations | 111 |
| Figure 3 | Long-term polyfactorial field experiment on the action and interaction of different crop rotations, soil tillage and fertilization | 111 |
| Figure 4 | Drawing the mouldboard plough at the experimental fields of Selectia Research Institute for Field Crops at Balti | 112 |
| Figure 5 | Laboratory building of the Selectia Experimental Station at Balti | 112 |
| Figure 6 | Field day at Selectia Experimental Station introduced by one of the authors, Prof. Boris Boincean | 113 |
| Figure 7 | Field day on harvesting of winter wheat at Selectia Experimental Station | 113 |
| Figure 8 | Most senior author, Prof. Igori Krupenikov with his grandson | 114 |

## Chapter 18

| | | |
|---|---|---|
| Figure 1 | Yields of winter wheat in Moldova, 1961–2001, projected to 2011 | 123 |
| Figure 2 | Yields of maize in Moldova, 1961–2001, projected to 2011 | 123 |
| Figure 3 | Crop yields in long-term field experiments at Balti, 1962–2002, projected to 2011: Yields of winter wheat at Selectia Experimental Station, rotation 2, 1962–2002 | 124 |
| Figure 4 | Crop yields in long-term field experiments at Balti, 1962–2002, projected to 2011: Yields of winter wheat at Selectia Experimental Station, rotation 4, 1962–2002 | 124 |
| Figure 5 | Crop yields in long-term field experiments at Balti, 1962–2002, projected to 2011: Yields of maize at Selectia Experimental Station, rotation 4, 1962–2002 | 124 |
| Figure 6 | Crop yields in long-term field experiments at Balti, 1962–2002, projected to 2011: Yields of continuous winter wheat without fertilizer at Selectia Experimental Station, 1965–2002 | 125 |
| Figure 7 | Crop yields in long-term field experiments at Balti, 1962–2002, projected to 2011: Yields of continuous winter wheat with fertilizer at Selectia Experimental Station, 1965–2002 | 125 |
| Figure 8 | Crop yields in long-term field experiments at Balti, 1962–2002, projected to 2011: Yields of continuous maize without fertilizer at Selectia Experimental Station, 1965–2002 | 125 |

Figure 9    Crop yields in long-term field experiments at Balti, 1962–2002, projected to 2011: Yields of continuous maize with fertilizer at Selectia Experimental Station, 1965–2002 . . . .   126

## Chapter 19

Figure 1    Distribution of humus, carbonates and clay in sediments exposed by the retreat of the Black Sea at Constanza; 1a Particle-size distribution; 1b Carbonates and humus . . . . . . . .   132
Figure 2    Humus and carbonate profiles of Trajan's Bank. 2a Trajan's Bank and buried profile; 2b Control natural soil profile . . . . . . . . . . . . . . . . . . . . . . . . . . . . . . . .   134

# List of Tables

## Chapter 3

| | | |
|---|---|---|
| Table 1 | Moldavian soils according to soil texture | 19 |
| Table 2 | Mean silt and clay content in chernozem | 20 |
| Table 3 | Particle size distribution of a *Typical chernozem* | 21 |
| Table 4 | Micro-aggregate size distribution of a *Typical chernozem* | 22 |
| Table 5 | Particle size distribution of a pristine *Common chernozem* | 22 |
| Table 6 | Micro-aggregate composition of a pristine *Common chernozem* | 23 |
| Table 7 | Comparison of micro-aggregate and particle size distribution of a cultivated *Typical chernozem* | 24 |

## Chapter 4

| | | |
|---|---|---|
| Table 1 | Mineral composition of heavy loam chernozem | 28 |
| Table 2 | Mineralogical composition of clay chernozem | 28 |
| Table 3 | Total elemental composition of light clay *Typical chernozem*, % of ignited, carbonate-free material | 29 |
| Table 4 | Total elemental composition of Moldavian chernozem, % ignited, carbonate-free material | 30 |
| Table 5 | Elemental composition of chernozem and silt, % ignited soil 0–50 (cm) | 31 |

## Chapter 5

| | | |
|---|---|---|
| Table 1 | Exchangeable bases in chernozem (mg/100 g soil) | 34 |
| Table 2 | pH in chernozem at 0–40 cm depth | 34 |

## Chapter 6

| | | |
|---|---|---|
| Table 1 | Soluble salt content of chernozem, mg/100 g of soil | 37 |

## Chapter 7

| | | |
|---|---|---|
| Table 1 | Humus balance (t/ha) and nutrient balance (kg/ha) in Moldavian soils | 42 |
| Table 2 | Humus and total nitrogen in heavy loam chernozem, % | 42 |
| Table 3 | Mean humus content in Moldavian soils | 43 |
| Table 4 | Relationships between soil humus and nitrogen and crop yields with forecasts to 2025 | 43 |
| Table 5 | Comparison of Dokuchaev's *Typical chernozem* with modern observations | 44 |
| Table 6 | Change in humus content of the plough layer (0–30 cm) of chernozem from the European USSR over 100 years | 45 |
| Table 7 | Comparison of soil organic carbon and nitrogen contents in *Typical chernozem* between native grassland, arable and bare fallow. After Mikhailova and others 2000 | 46 |
| Table 8 | Humus composition of Moldavian soils | 47 |

## Chapter 8

| | | |
|---|---|---|
| Table 1 | Nitrates in Moldavian soils, kg/ha | 54 |
| Table 2 | Nitrate, kg/ha, in soil layers to 10–20 m | 54 |

## Chapter 9

| | | |
|---|---|---|
| Table 1 | Phosphorus reserves (total of Chiricov's fractions) in Moldavian chernozem, mg/100 g of soil | 58 |
| Table 2 | Phosphate fractions in the plough layer of Moldavian chernozem, mg/100 g of soil | 59 |
| Table 3 | Reserves of the organic and mineral phosphorus in the 0–50 cm layer | 59 |
| Table 4 | Provision of mobile phosphorus in arable soils in Moldova, 1971–1990 | 59 |
| Table 5 | Mineral fertilizer use in Moldova, 1961–2002 | 61 |

## Chapter 10

| | | |
|---|---|---|
| Table 1 | Data on crotovinas in chernozem | 64 |
| Table 2 | Mesofauna, micro-organisms and respiration in the 0–30 cm layer | 65 |

## Chapter 11

| | | |
|---|---|---|
| Table 1 | Physical properties of Moldavian chernozem | 71 |
| Table 2 | Plant-available water in chernozem | 72 |
| Table 3 | Comparison of structure of *Calcareous chernozem* and *Xerophyte-wooded chernozem* under forest and ploughland | 73 |
| Table 4 | Grouping of soils according to agrophysical properties | 74 |
| Table 5 | Moisture conditions of the growing season in Moldova | 74 |

## Chapter 12

| | | |
|---|---|---|
| Table 1 | *Xerophyte-wooded chernozem* roots and humus content, t/ha | 85 |
| Table 2 | Altitudinal distribution of chernozem | 86 |
| Table 3 | *Bonitet* rating of the main soil types of Moldova | 87 |

## Chapter 15

| | | |
|---|---|---|
| Table 1 | Distinctions between natural ecosystems and agroecosystems | 100 |
| Table 2 | Energy balance of long-term field experiments at Selectia Experimental Station, Balti, 1962–1991 | 100 |
| Table 3 | Mineralization of labile soil organic matter under crop rotations at Selectia Experimental Station, Balti, 1993 | 101 |
| Table 4 | Mineralization of soil organic matter (SOM) at Selectia Experimental Station, 1982–1991, mean values | 102 |
| Table 5 | Ash content of plants, mean % total mass | 103 |

## Chapter 17

| | | |
|---|---|---|
| Table 1 | Changes in soil organic matter stocks in the 0–20 cm layer under different arable rotations, monocultures, bare fallow and meadow at Selectia Experimental Station, 1962–1991 | 115 |
| Table 2 | Fertilizer efficiency in crop rotations and monocultures in long-term experiments at Selectia Experimental Station, 1991–1996 mean yields, t/ha | 117 |

## Chapter 18

| | | |
|---|---|---|
| Table 1 | Trends of winter wheat and maize yields in Moldova and in long-term field experiments at Selectia Experimental Station, 1962–2002 | 123 |
| Table 2 | Soil organic matter content (%) in the polyfactorial experiment at Selectia Experimental Station | 127 |

| Table 3 | Yields of winter wheat and sugar beet in the long-term polyfactorial field experiment at Selectia Experimental Station, 1996–2009, t/ha 1 – winter wheat; 2 – sugar beet | 128 |

## Chapter 19

| Table 1 | Structure of humus in the Trajan's Bank soils | 134 |
| Table 2 | Aggregate-size distribution in the Trajan's Bank soil and buried profile 250 K | 135 |

# Part I
# The State of the Land and Its Management

# Introduction

**Abstract** In common with much of eastern and southern Europe, Moldova faces poverty, agricultural crisis, a flight from the land, and degradation of the soil. And yet, the black earth, or chernozem, remains amazingly resilient – thanks to its unique physical and mineralogical inheritance, thickness of topsoil and the way that it repairs itself like a living thing. Ecological agriculture can build on these attributes to maintain the soil's resilience and increase its productivity. The following chapters highlight key questions of land use and the fundamental attributes of the chernozem that help, or hinder, their protection and restoration. Finally, we explore ways and means of accomplishing the much-needed conversion of agriculture to the ecological way. In the fifth century BC, Herodotus described the country between the Prut and Dniester when the virgin steppe was ploughed for wheat and wine, much of it for export to metropolitan Greece. Waves of nomadic tribes passed through during the first century AD and, again, at the time of the emergence of Moldavia as a feudal state but iron implements from the fourteenth century attest to highly developed agriculture and animal husbandry. The land was laid waste again by the Turks but travellers remarked "Although the country is ravaged, there is no other place with such soils. The Promised Land: grows everything!" and Cantemir's *Descriptio Moldaviae* in 1715 records "The Moldavian fields are known from the oldest writers' works as very yielding, excelling the treasure of the mountains." The term chernozem was introduced into the scientific arena by Lomonosov in 1765. Various theories on its genesis were advanced: marine, marshy, and terrestrial. Only at the end of the nineteenth century was it generally accepted that the chernozem is a soil – after Dokuchaev visited the country in 1877, collecting material for his book *Russian Chernozem* which settled the issue and founded soil science. Under the Soviet Union from 1945, Dimo ushered in the modern period with detailed soil surveys and an array of fundamental information was compiled over the following decades. But the era of modern science coincided with the onset of a more exploitative farming system; the long grass fallow system at the turn of the nineteenth century gave way to intensive row crops and a consumerist attitude – exploiting the soil without taking the trouble to conserve it as a resource. The worldwide success of intensive row-crop systems was described at the time as the green revolution but the yields achieved through widespread application of fertilizers and broad-spectrum pesticides masked, for a time, a real decline in soil organic matter and soil health. Now this must be addressed.

**Keywords** Chernozem · Ecological agriculture · Historical land use and management · Land degradation · Soil fertility

## In Essence

Moldova is in crisis – facing poverty, a flight from the land and, at the most fundamental level, degradation of the land and soil. Compared with 20 or 30 years ago, agriculture too is in decline – both because of the loss of soil and soil fertility and, also, because of the economic and political climate. In 1992, the Rio summit on sustainable development drew the world's attention to the imperatives of conserving soil, biodiversity and habitat, proclaiming "Economic development out of touch with ecology leads to desertification" (WCED 1987). At that time, Moldova was an independent state;

its delegation participated in the summit and raised no objection to its findings. Yet, 20 years on, we see more and more land degradation in all its forms. *Soil Degradation and Desertification*, published in 2000 (Ursu 2000), reports a 1% annual erosion of arable land. Except for some islands of good practice, farming is estranged from its scientific basis; crop yields and total production are declining, and the country's external debt is hardly supportable.

Moldova is not alone. Neighbouring countries are in much the same plight but, unlike Russia, Moldova cannot make good its economic losses by exporting oil and gas. Even in Russia, scientists are alarmed by the state of their soils (Bondarev 2002, Parahnevici and Parahnevici 2002). In the proceedings of the conference in 2002 celebrating the 75th anniversary of the institute named after V.V. Dokuchaev, the word *sustainable* appeared in the titles of 240 out of 400 contributions from Russia and neighbouring countries dealing with the constitution and fertility of soils in agriculture, forestry and, even, recreation.

We would do well to ponder on the nature and uniqueness of our black earth, known to science by its Russian name *chernozem*: not only its native fertility but also its susceptibility to loss of humus and nutrients, to erosion and compaction under cultivation and to salinity, sodicity, waterlogging and chemical contamination. In spite of all this, it remains amazingly resilient – thanks to its unique physical and mineralogical inheritance, the thickness of the topsoil and the way that it repairs and rebuilds itself like a living thing.

We can build on these attributes so as to maintain the soil's resilience and increase its productivity. This is what we are calling *ecological agriculture*. Best practice includes minimum tillage – to avoid compaction and accelerated loss of soil organic matter and, at the same time, use less energy; it respects crop rotation that includes leguminous crops; builds up soil organic matter by returning crop residues, straw, manure and compost; employs soil conservation measures, especially on sloping ground, and includes prudent restoration of irrigation and drainage works. Ecological agriculture does not shun the optimum use of mineral fertilizer and, when appropriate, chemical control of pests, diseases and weeds – but it does eschew damaging practices such as straw burning, spring ploughing, cultivation up- and downslope and salinization by over-irrigation and poor-quality water; and it takes care to avoid the contamination of soil and water. Appropriate local land improvement measures include reclamation of solonetz, landslides and gullies and restoration of eroded soils with natural materials such as peat, muck and topsoil taken from construction sites.

It is easy to propose but hard to accomplish all this against political opposition, financial constraints and general ignorance of the condition of the soil. But, as a country, we have no choice: as industrial production falls, agriculture has to carry an increasing proportion of the economy (now 70–80%). Moldova has become, once again, an agricultural country – so some order is necessary in our agricultural affairs. The essential change from soil exploitation to soil conservation – to ecological agriculture – has been blocked by the destructive consequences of the, so-called, *Land* program which broke up the established structure of farming and depressed traditional specialisms like plant breeding, stockbreeding, viticulture, horticulture and production of tobacco and oil seeds. The redistribution of land holdings and production quotas violated both the law of the land and the most elementary rules of soil management; and financing this program has made Moldova the poorest country in Europe.

How to put things right? Nowadays, there are moves to consolidate land holdings that might have been expected long ago, in view of economic necessity and the natural management units of the landscape. Surely, this is one of the keys to land improvement and agricultural efficiency, and a lesson that we can learn from the experience of the USA in recovering from the simultaneous catastrophes of the Great Depression and the Dust Bowl. In the 1920s and 1930s, America worked its way through the exhaustion of its soils and its farmers to an accommodation with the new situation. By 1940, two and a half million farmers had left the Great Plains; some farms were enlarged and many more, especially the small ones, disappeared. Amongst a raft of initiatives to tackle immediate and longer term social and financial problems, the scientific basis for this remarkable recovery was laid by the establishment of a national Soil Conservation Service, backed up by a nationwide soil survey which has provided detailed soil information for the whole country. Governments across Europe should, likewise, adopt a *Soil Law* and create a properly funded soil service to implement it.

Land and soil are finite. We cannot take more out of them than we put in – we reap what we sow. The following chapters highlight key questions of land use and the fundamental attributes of soils that help, or hinder, their protection and restoration. Finally, we explore ways and means of accomplishing the much-needed conversion of agriculture to the ecological way. In this task, we draw upon the symbiosis of pedological and agricultural knowledge begun more than a century ago by V.V. Dokuchaev and V.R. Viliams.

## A Short History

A few words about the history of the knowledge of our soils may set the scene for a discussion about the relationship between soil science and agriculture. Dokuchaev formulated the scientific understanding of soil in Bessarabia in the last quarter of the nineteenth century but our story begins 15 centuries earlier, and the chain of empirical knowledge has been maintained through all the events of history because of the uniqueness of the soil and the multifarious interests of many and various authors.

The earliest surviving mention of the country between the Prut and the Dniester comes from the Greek, Herodotus, in the fifth century BC. In his *History*, he describes the region, called *Little Skifia*, as "A plain with thick black soil, rich in grass". The rivers had "turbid water", which suggests soil erosion even then. Greeks settled the lower Dniester in the seventh century BC, calling the colony *Tira* and the river *Tiras* – it was navigable and rich in fish. The *Tirita* people grew wheat and wine; grain was exported to metropolitan Greece, and fish and bunches of grapes are pictured on their coins.[1] And so the steppe was ploughed.

At the beginning of the first millennium of the Common Era, the steppe character of the region was described by Strabo in his *Geography*; he was followed in the first century by Roman authors such as Pliny the Elder and Flac, in the fourth century by Ammian Marstellin and in the sixth century by the Gothic historian Iordan.

The marauding Lithuanian Duke Iagalio, reaching Mounkastro (Belgorod-Dnestrovskiy in the Odessa Region) in 1424, said the country "is yielding and is a fruit-bearing land". In the sixteenth century, the Polish historian Goretski described it as "abundant in barley and wheat"; Stefan the Great's Venetian doctor, Matei Murano, wrote in 1502 "this country is yielding, pleasant, and rich in cattle and various fruits, the pastures are excellent"; and in 1541 the Hungarian King Reihercdorf's adviser wrote "it is poor in arms, proud of its deeds, rich and fertile". All the land was not under the plough but even the heaviest soils might be ploughed – the Italian, Gratiani, remarked "... sometimes 6 pairs of bullocks are harnessed".

A century later, another Italian, Mark Bandini, wrote in 1648 "the yields are high but, until you see it, it is difficult to believe the abundance of Moldavian lands"; and Charles de Jopecourt, traversing the country in 1611–1617, recorded "There is a lot of wheat, rye, barley, oats, hay... we can see many slopes rich in grapes" (the same was reported by other travellers of the period). He remarked on the diversity and the productivity of the arable crops but concluded "the greatest treasure of Moldavia is the vineyards"; wine was exported.

Interest in the country increased with the Russo-Turkish wars at the end of the eighteenth and beginning of the nineteenth century. In 1811, Talleyrand advised Napoleon not to cede Bessarabia because "its soils are excellent and will be profitable". Czar Alexander's subsequent manifest on its annexation to Russia states: "it represents yielding ground", which harks back to Cantemir and, even, earlier. In *Lands of Bessarabia*, Andrew Mayer in 1794 noted that the *power* of the soil changed from "light and (shallow) to dark and deep" and, near the Black Sea, a "chernozem stratum" appeared. The term *chernozem* was introduced into the scientific arena by M. Lomonosov in 1765 and Mayer's reference was connected with the long debate about its genesis. Various theories were advanced: marine, marshy, and terrestrial. Only at the end of the nineteenth century was it generally accepted that the chernozem is a soil but, P. Svinina writes, in 1816, in a little monograph *Description of the Bessarabia Region* "The soil in this region consists only of chernozem that lies in a hollow of 2 ft, and some places so deep that the ploughman would never touch the bottom ... This kind of soil appeared as a result of rotting of unused plant residues, which formed this nice soil." A naïve but accurate explanation, half a century before Rupprecht and Dokuchaev!

---

[1] As described by T. Passeka in *The Early Agricultural Tribes from Podnestrovia* (Krupenikov 1999).

The chernozem was becoming a scientific issue. I. Saburov, in 1826, commented that the soil of northern Bessarabia was fertile, rich in nitrogen, held abundant moisture for the crops and was excellent for grain production whereas, in the southern part, the soil was poorer and yields less reliable owing to drought. Further information about the soils is found in academician Afelda's geography, the work of the scientist K. Arseniev in 1848, and even the statesman Tengoborski in 1885. Now it became important to formulate a correct definition of the term *chernozem* and to understand the genesis, distribution and fertility of the soil and the difference in productivity between the north and the south. The Moldavian agronomist A. Grosul-Tolstoy, in 1856, wrote *A View of the Rivers, Soils and Agricultural Situation of the New Russian Region of Bessarabia*. He included a zonal map of the soils from the Prut to the Ingul (modern Moldova and a part of the Ukraine) showing in sequence, from north to south, the *real chernozem, sandy loam chernozem, clay loam with included chernozem* and *calcareous clayey*. Like Lomonosov, he used *chernozem* in two senses, both as a soil and as the unique kind of humus but, in spite of imperfect definition, the distinctive soil zones were mapped for the first time. He also reported on grain yields, droughts and, again for the first time, soil erosion. The enthusiastic opinion of the geographer and ethnographer Afanasieva Tchujbinski rounds off this period of knowledge development in 1863: "Grain production, with the soils' generosity, can flourish here; viticulture is the best; and tobacco cultivation needs only some small improvements to transform our tobacco trade" – his words are still apposite.

A new era began when Dokuchaev visited the country in 1877, collecting material for his book *Russian Chernozem* (Dokuchaev 1883, 1952) which settled the issue of the genesis of the chernozem and founded pedology. His first account is a concise statement of the principles of a new science – flattering to us and cited in many subsequent publications. He called the chernozem of the Balti steppe *first class* and typical in its structure, thickness and humification. Some years later, he undertook a thorough analysis of the soils of Bessarabia which he called "the most interesting province in European Russia": he categorized the northern and southern chernozem as *typical* and *transitional*; reported on saline soils and eroded soils; and observed the topographic zonation of soils – especially in Codri: "Bessarabian Switzerland". Out of these observations came a mature theory of soil genesis and a scheme of classification that remains fundamental and recognizable today.

By the turn of the century, Dokuchaev's ideas were elaborated and applied further afield, especially by N.M. Sibirtsev. Nicolai Dimo, native of Moldavia and pupil of Dokuchaev and Sibirtsev, researched in the provinces of Saratov and Penza. The Bessarabian, Zemstvo, began work in 1913 with A. Pankov and others; despite the dislocation of world war, Pankov produced what he called a tentative provincial map showing the sub-types of the chernozem[2] and, like Dokuchaev, he showed the best chernozem in the Balti steppe. Other research included Michael Karchevski's *Materials on the Study of Natural–Historical Conditions of Bessarabian Viticulture* in 1918, which distinguished micro-regions that may still be used (for instance, the light soils in Codri, favoured for vines). I. Nabokih distinguished various chernozems in Bessarabia, saline soils in the valley of the Reut and in the northern-western periphery of Codri and made a collection of soil monoliths that is still used in the Odessa Agricultural University.

From this time, we may discern synergy between agronomy and pedology. An experimental station, planned by P. Kosticev, was established on *Typical chernozem* in the Trubetski's Provinces property near Ribnita Plotianskaia. A nine-field crop rotation was instituted, experiments were conducted on tillage and manuring and the effectiveness of phosphate fertilizers was demonstrated by Karabetov in 1905 (Ploteansk Agricultural Experimental Station 1913) and Velibeli in 1914. Further fertilizer and tillage experiments, especially tillage to counter soil erosion, were conducted in Cocorozeni and Grinaut by G. Kozlovski in the first decade of the century; these reports can still be used for comparison. At the end of this period, L. Berg's *Bessarabia: The Country, People, Husbandry*, published in Petrograd in 1918, reported "This country, with a beneficent climate, allows cultivation of such valuable crops as grains, tobacco, corn, with fertile soils providing 120 million pounds of wheat per year."

During the period when Bessarabia was a part of Romania (1918–1940), it attracted the attention of

---

[2] Pankov's manuscript map is archived in the Dokuchaev Institute in Moscow.

George Murgoch's school of pedology, which followed Dokuchaev and Sibirtsev.[3] He mapped the richest chernozem (containing up to 8% humus) in the Balti steppe; in the south, the *Codri chocolate chernozem* (Sibirtsev's terminology) was shown to be equivalent to the modern *Common chernozem*; forest soils were divided into *podzolic* (*Grey* nowadays) and *red-brown* (*Brown* nowadays). A soil map of Bessarabia at scale 1:25 million, after Murgoch, is shown in the S*tatistical Album Atlas* published in French in Bucharest in 1929, and T. Zaidelea's map at scale 1:1.5 million, published in 1927, is close to Murgoch's map (Krupenikov 1999). These maps were important scientifically but, of course, too small scale to be of practical application. Larger scale maps appeared in the 1920s and 1930s: N. Florov made maps of Lapushna District at scale 1:130,000, Soroca District at scale 1:1 million and Kopanca village at scale 1:30,000 but only the last found any application in agricultural practice. In fact, Dimo was right to observe that, during the 1940s, the soils of Moldavia were mapped and analysed in less detail than those of the Russian tundra! The reasons for this included two revolutions, two world wars and the ambiguous political position of Bessarabia.

Under the Soviet Union from 1945, Dimo ushered in the modern period with detailed soil surveys. Building on this foundation, A. Ursu published an array of fundamental information in his three-volume *Moldavian Soils* (Ursu 1984, 1985, 1986), and many general works on agriculture include those by Sidorov (1965, 1989), P. Kibasov and others (1984) and Boincean (1999).

Agriculture is the practical application of knowledge of the land, including soil science, to maintain and increase its productivity. Sidorov's stages of development of farming systems in Moldavia reflect the evolution of agriculture in Moldova:

- Grass fallow with arable break crops in the 1880s and 1890s
- The fallow system at the turn of the nineteenth century
- Continuous arable with row crops and intensive chemical agriculture from 1954

The intensive row crop system achieved early and almost worldwide success as the *green revolution* but yields levelled off and, now, yields are even declining and the negative impacts of chemical-dependent farming are evident in pollution of the environment and human health issues. Certainly in Moldova, success was an illusion. The agricultural scene is strewn with great plans: for specialization and concentration of farm production, big processing factories, subsidy of the farm sector by cheap non-renewable energy, and controlled farm-gate prices for agricultural products. The disparity between the real costs of energy and machinery and the low farm-gate prices for the agricultural products was aggravated by the disintegration of the Soviet Union; now we face a harsh adjustment to a very different economy of limited and expensive agricultural inputs.

The pressing need for an alternative farming system throws us back upon something very like the *plodosmen* cereal-grasses system developed by Dokuchaev, Kostichev, Timireazev and Viliams and epitomized by

- Organisation of the management of the land according to its natural landscapes
- Crop rotation including perennial crops
- Growing of the crops and varieties that are adapted to each particular region
- Land improvement by appropriate application of fertilizers, drainage and other techniques
- Prudent cultivation – with adaptive measures, such as shelter belts and farm ponds, to soften the negative effects of drought, unfavourable winter conditions and erosion

## References

Boincean BP 1999 *Ecological farming in the Republic of Moldova (crop rotation and soil organic matter)*. Stiinta, Chisinau 269p (Russian)

Bondarev AG 2002 On soil resilience and sensitivity to compaction by agricultural implements. 208–209 in NB Hitrov (editor) *Soil resilience to natural and human influences*. Nauka, Moscow (Russian)

Dokuchaev VV 1883 *Russian chernozem*. Independent Society for Economics, St Petersburg (Russian)

---

[3] The Romanian school's continuing interest in chernozem is echoed in the Western European literature by Ghitulescu and Stoops (1970), Ghitulescu (1971) and Pons-Ghitulescu (1986).

Dokuchaev VV 1952 *Russian chernozem. The report of the independent economics society, second edition with foreword by VR Viliams*. State Publisher of Agricultural Literature, Moscow 634p (Russian, English translation 1967 Israeli Program for Scientific Translations, Jerusalem)

Ghitulescu N and G Stoops 1970 Etude micromorphologique de l'activité biologique dans quelques sols de la Dubroudja-Roumanie (Vermustoll pachique – Chernozem chocolat dans la classification Roumaine). *Pedologie (Ghent)* 3, 339–356 (French)

Ghitulescu N 1971 Etude micromorphologique de quelque sols de la Plaine de Cilistea – Roumanie (Vermudoll haplique – Chernozem fortement lévige dans la classification Roumaine). *Pedologie (Ghent)* 7, 131–151 (French)

Kibasov PT, IS Constantinov, IS Constantinov and IA Turucalo 1984 *Crop rotations for intensive agriculture*. Cartea Moldoveneasca, Chisinev 194p (Russian)

Krupenikov IA 1999 Twenty five centuries of soil knowledge in Moldova (from Herodotus to Dimo). in *Soil science in the Republic of Moldova at the end of the second millennium*. Stiinta, Chisinau (Russian)

Parahnevici TM and MI Parahnevici 2002 Agrophysical degradation of *Leached chernozem* and how to reduce it. 128–129 in NB Hitrov (editor) *Soil resilience to natural and anthropogenic influence*. Nauka, Moscow (Russian)

Ploteansk Agricultural Experimental Station (1913) *Report for 1912*. 380p. Odessa (Russian)

Pons-Ghitulescu N 1986 *Chernozem calcique, Roumanie*. Soil monolith paper 7, ISRIC, Wageningen 83p (French)

Sidorov MI 1965 *The System of agriculture in Moldova*. Cartea Moldoveneasca, Chisinau 170p [and new edition 1989] (Russian)

Ursu AT (editor) 1984 *Soils of Moldova. Vol. 1 Genesis, ecology, classification and systematic description of soils*. Stiinta, Chisinau, 351p (Russian)

Ursu AT (editor) 1985 *Soils of Moldova. Vol. 2 Soil geography, description of soil provinces, districts and sub-districts*. Stiinta, Chisinau, 239p (Russian)

UrsuAT (editor) 1986 *Soils of Moldova. Vol. 3 Soil management, protection and improvement*. Stiinta, Chisinau, 333p (Russian)

Ursu AT (editor) 2000 *Soil degradation and desertification*. Academy of Sciences of Moldova, Chisinau, 307p (Romanian)

WCED 1987 *Our common future*. World Commission on Environment and Development. Oxford University Press, Oxford

# Part II
# The Anatomy of the Black Earth

# The Soil Cover

**Abstract** We review the features of the soil cover that determine its biological productivity, its resilience in the face of disturbances and its agricultural utilization – especially by ecological agriculture. We focus on the black earth, or chernozem, drawing upon an extensive literature and recent approaches to the understanding of the soil as the Earth's living skin. Worldwide, chernozem soils comprise a tract of 240 million ha of the middle-latitude steppe of Eurasia and prairies of North America, constituting some of the world's main grain-producing regions. In Moldova, chernozem occupy 75–80% of the country and the various sub-types exhibit a clear north–south zonation: from *Podzolized* to *Leached, Typical, Common* and, finally, *Calcareous chernozem*. Terminology follows the 1987 soil classification of Moldova (Krupenikov and Podimov 1987) which categorizes soils according to their morphology; the *World Reference Base for Soil Resources* is used for international correlation. Dokuchaev and Vernadsky perceived soils as independent, living, natural bodies – and soil science as a natural science in its own right. As the interface between the atmosphere and lithosphere, the soil creates local conditions for an immense variety of physical, chemical and biological processes. Soils exhibit a sequence of levels of organization from mineral crystals and organic molecules to the organo-mineral complex, to soil aggregates, horizons and profiles, to catenas, landscapes and, finally, the whole pedosphere. Agriculture influences the soil directly at the level of the organo-mineral complex and at the level of the aggregate; we plan and check upon the outcomes of management at the soil profile and catena levels and establish indices of performance within the limits of landscapes; the pedosphere is the object of interest and care of mankind. The sustainability of soil can be considered as its ability to maintain its state under variations of soil-forming factors and, provided the topsoil is preserved, reconstruct itself after perturbation. Particle-size distribution and chemical composition, inherited from the parent material, are stable under cultivation but not so humus – which must be maintained by judicious management practices. On sloping ground, intense summer rainstorms combined with inadequate management have brought extensive and, sometimes, severe soil erosion that might be prevented by respecting crop rotation and maintaining good ground cover. Chernozem remain productive, even with half the original profile gone, but eroded soils should be farmed only by maintaining the whole arsenal of conservation-farming techniques.

**Keywords** Black earth/chernozem · Soil geography · Soil organization · Sustainability

## The Pattern of Soil Cover

Worldwide, chernozem extend in a broad belt of some 240 million ha of the middle-latitude Eurasian steppes and North American prairies (Fig. 1). The neighbouring chestnut-coloured earths or *Kastanozem* of the drier, short-grass steppe, similar except for a lower humus content and still potentially fertile land, extend over 465 million ha in southern Ukraine and Russia, Kazakhstan, Canada, the USA and Mexico and in the Pampas and Chaco of South America.

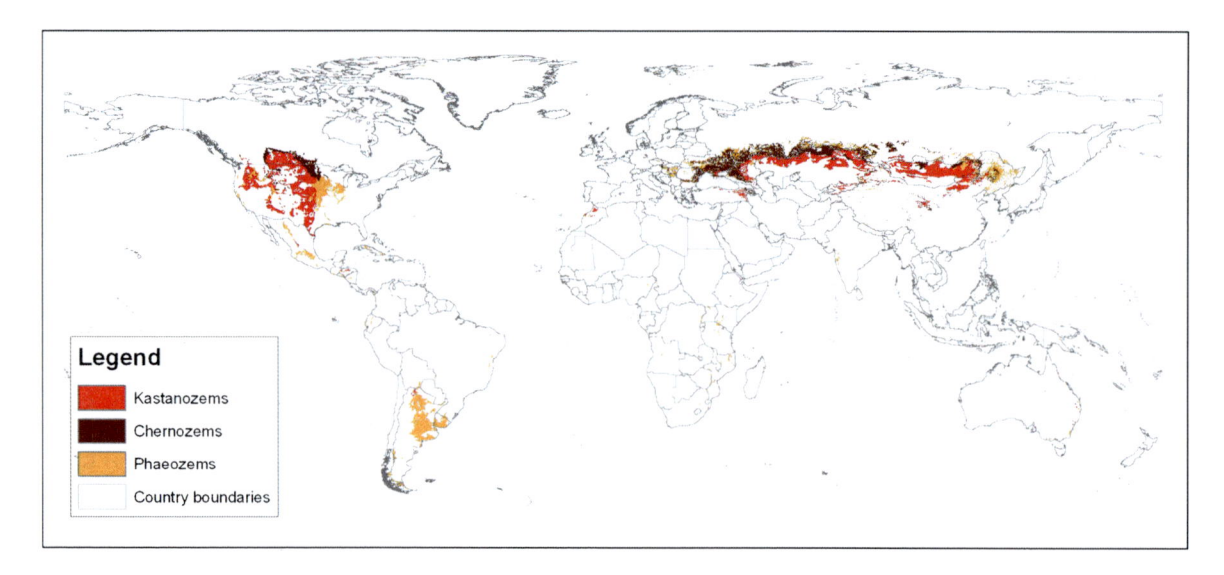

**Fig. 1** Global distribution of chernozem and related soils. Prepared by FAO from the Digital Soil Map of the World

In Moldova, *Typical chernozem* occupy some 817,000 ha. By *typical*, we mean that it represents the central concept of the chernozem[1]; it is the climax of soil development under steppe vegetation and the gold standard in terms of land quality. There are also large areas of *Leached chernozem*, *Common chernozem* and *Calcareous chernozem*; the various sub-types occur in distinctive latitudinal zones that are interrupted by high ground, which shows its own altitudinal sequence of soils (Ursu 1985). Most of our soil maps, from detailed surveys at scale 1:10,000 to compilations at scale 1:400,000, are in manuscript and accessible only in archives. However, quantitative information for the areas under different soil types is published in *Soils of Moldova* (Ursu 1985). Excluding soils occupying only small areas, *Chernozem* and related soils occupy 75–80% of the country, *Grey* and *Brown forest soils* about 12% and the wet or seasonally wet *gley soils* of the alluvial plains about 8%. Figures 2, 3, 4 and 5 illustrate a *Typical chernozem*, an example from the Kursk region of Russia; *Dark grey forest soil* of the adjacent forest zone; the *Kastanozem* of the drier steppe regions to the south, an example from Ukraine, and *Gley* soils that occur in water-receiving depressions within the black earth region.

---

[1]*Haplic chernozem* in the terminology of the *World Reference Base for Soil Resources 2006* (IUSS WRB 2006).

**Fig. 2** *Typical chernozem* at the V.V. Alekhin Central-Chernozem Biosphere State Reserve near Kursk, Russia. David Dent

**Fig. 3** *Kastanozem* in the short-grass steppe of southern Ukraine. David Dent

**Fig. 4** *Dark grey forest soil* in the Russian forest-steppe zone. David Dent

**Fig. 5** *Gley soil* in a broad depression within the short-grass steppe of southern Russia. David Dent

All the Moldavian chernozem are considered *thick chernozem*; they contain at least 1% of humus at a depth of 1 m and there is an exotic variety in the pedunculate oak (*Quercus robur*) forests with an extraordinary 12% of humus in the topsoil, very strong soil structure and a remarkably diverse and numerous population of earthworms and other invertebrates.[2]

In terms of reaction, there is acidification in *Leached chernozem* and *Podzolized chernozem*, which are transitional to *Grey forest soils*. At the other end of the scale, *Calcareous chernozem*[3] are unique in many aspects in Moldova and, more generally, in the Danube–Pontic region of Europe (Krupenikov 1979, Sinkevici 1989, Alekseev 1999) and these soils need some special approaches and techniques for sustainable use. The various chernozem sub-types exhibit a clear north–south zonation: from *Podzolized* to *Leached, Typical, Common* and, finally, *Calcareous chernozem*; they are described in detail in Chapter 12. Also within the chernozem zone, we find inclusions of some 30,000 ha of *Solonetz* which are sodic (a high content of exchangeable sodium that renders them strongly alkaline) but which may be ameliorated by a combination of gypsum applications and leaching (Ursu 1986).

*Grey forest soils*, characterized by grey topsoil over a clay pan, occur in the north of the country, in Codri and the Transdniestrian Heights. *Brown forest soils*, which are brown throughout and lack any marked textural contrast in the soil profile, are restricted to a small area of about 27,000 ha in the high part of Codri. Nowadays, only about half of the so-called forest soils are under forest – where their natural profiles remain intact; under arable and vineyards, they easily lose humus and are exposed to erosion.

Periodically flooded soils developed in alluvium are a diverse group. *Meadow soils*, if protected from flooding, are fertile and good for fruit and vegetable crops. On the other hand, there are extensive tracts of *saline soils*, *swamp*, and *compact soils* – commonly juxtaposed in the bottomlands of the small rivers and along the lower reaches of the Prut. Formerly, some bottomlands were ameliorated, though not always successfully; nowadays they are mostly neglected, rough grazing.

Intense summer rainstorms combined with hilly terrain and inadequate management can easily result in severe soil erosion. Unfortunately, in Moldova, the rules of conservation farming are not well established or well adopted. It is hard to compute the extent of eroded areas because the old soil surveys were not standardized and modern re-survey is slow work. However, the principal monograph on soil erosion in Moldova (Nour 2001) estimates that 30% of the country has suffered some degree of soil erosion: 900,000 ha is categorized as *slight*, 250,000 ha as *medium* and 200,000 ha as *severe*. The precise numbers are disputable but there can be no doubt that land degradation is approaching critical limits with erosion on arable and steep land, expanding gullies and the spread of salinity and waterlogging.

We may conclude from this brief overview that

- On the plains and gentle slopes (up to 2°), land that retains an intact soil profile[4] can be used sustainably for productive, ecological agriculture – which means respecting crop rotation, maintaining good ground cover and judicious application of fertilizers and manure to maintain nutrient status, soil organic matter and soil structure.
- Eroded soils, and especially the medium and severely eroded soils, should be farmed only by strictly maintaining the whole arsenal of conservation-farming techniques: eschewing row crops, respecting biological conservation measures, cultivating and sowing on the contour, intercepting runoff so that it can infiltrate into the soil and even rebuilding the soil profile by bringing in soil material. Nor should the slightly eroded soils be taken for granted; they have already lost about half of their topsoil but, while they still qualify as chernozem, ecological agriculture is still sustainable – chernozem remain productive even with half of the original topsoil gone but we must guarantee soil protection from now on.

---

[2] *Xerophyte-wooded chernozem* or *Vermic chernozem* according to WRB.

[3] *Calcic chernozem* in WRB.

[4] The term soil profile describes the soil's natural vertical sequence of topsoil and subsoil layers, as seen on a pit or quarry face and as illustrated in Figs. 2, 3, 4, and 5.

- Swamps and bottomlands should be used only in special projects involving drainage, leaching of excess salt and amelioration of sodicity using gypsum. Even then, crops must be tolerant of wetness and salinity (Krupenikov 2001, Krupenikov et al. 2002).

## Sui Generis

Dokuchaev perceived soils as individual natural bodies, "the 4th kingdom of nature", in a league with minerals, plants and animals. In his pedological paradigm, soil has its own space, occupied only by itself, and its own time, which is faster than the geological time but slower than human history. Soil science, sui generis, is a natural science in its own right – not a part of geology, biology or agronomy but a peer alongside other natural sciences.

This concept was elaborated by Dokuchaev's student and friend V.I. Vernadsky who described soil as "an alive substance" (Krupenikov 1988). In his own words: "There is not a chemical power in continuous development on the Earth's surface and, therefore, a more powerful one in its final consequences, than a living organism" (Vernadsky 1965, p. 241). He perceived the biosphere encompassing not only living organisms but also the habitat and products of life, their functions and influence over geological history. Its components are living organisms, soils, the oceans, surface and groundwaters – full of life and products of its activity; and the atmosphere – in which one of the main components, oxygen, accumulated as a result of the photosynthesis of the green plants. Even the Earth's crust beneath the soil exhibits ancient manifestations of life, especially limestone, rock phosphate, coal and lignite that are the direct products of living organisms.

Soil is the interface between the living and the inert, between the Earth's crust, the atmosphere and the hydrosphere – it embraces living organisms and organic matter, all rich in chemical energy. Indeed, soil is the domain of the highest geochemical energy and activity; the upper 1 or 2 m of soil surpasses the biological activity and geochemical energy of the much greater vertical extent of the atmosphere and hydrosphere. For instance, a few centimetres below the surface of underwater clay, lack of oxygen greatly restricts biochemical processes. Hence Vernadsky's contention: "The significance of soil on our planet's history is bigger than it appears. The soil ... determines the throughflow of water from rainfall ... thus the composition of the fresh water is ... directly determined by the soil chemistry." Even the chemistry of river water "is caused by the chemical work of soil" (Vernadsky 1967, p. 342). Unfortunately, this also leads to the pollution of surface and groundwater with nitrates, pesticides and other toxic substances.

The idea of soil as an independent natural body, well understood by Dokuchaev and Vernadsky, somehow offends some biologists. Living organisms are specified by their genetic code but soil does not have a genetic code; it develops according to its own laws. Soil is both biological and mineral; it is dynamic but, at the same time, there are limits to its composition and architecture that define its behaviour. It has constant relationships with other forces of nature that we call the *factors of soil formation*[5]: climate (atmosphere and hydrosphere), parent material (lithosphere), living organisms (biosphere in the narrow sense of the word), topography (which redistributes materials and energy) and time (the period of soil formation). If agriculture is to be productive and sustainable, it needs to take account of all these factors and provide for the soil's own biological diversity, functions and productivity.

In a modern context, a distinguished American Committee on Opportunities in Soil Science (Sposito and Reginato 1992) depicts the soil as the frame of life: "Some suggestions appeared that life didn't arise in a water medium but in soil, and the clay minerals could play the role of matrix in the synthesis of biological molecules." We usually think that there is no soil without organic matter – but clay minerals may form abiotically and the evolution of soil and the soil cover may be conceived as a sequence of levels of organization: crystalline minerals and organic molecules → organo-mineral complex → aggregate → soil horizon → soil profile (or its three-dimensional manifestation, the pedon) → catena → soil landscape → pedosphere (Fig. 6).

This is not just theory. In practice, agriculture influences the soil at the level of the organo-mineral complex and directly at the level of the aggregate. The soil

---

[5]Identified by Dokuchaev although first written in the modern form by Hans Jenny (1941).

**Fig. 6** Levels of soil organization (after Sposito and Reginato)

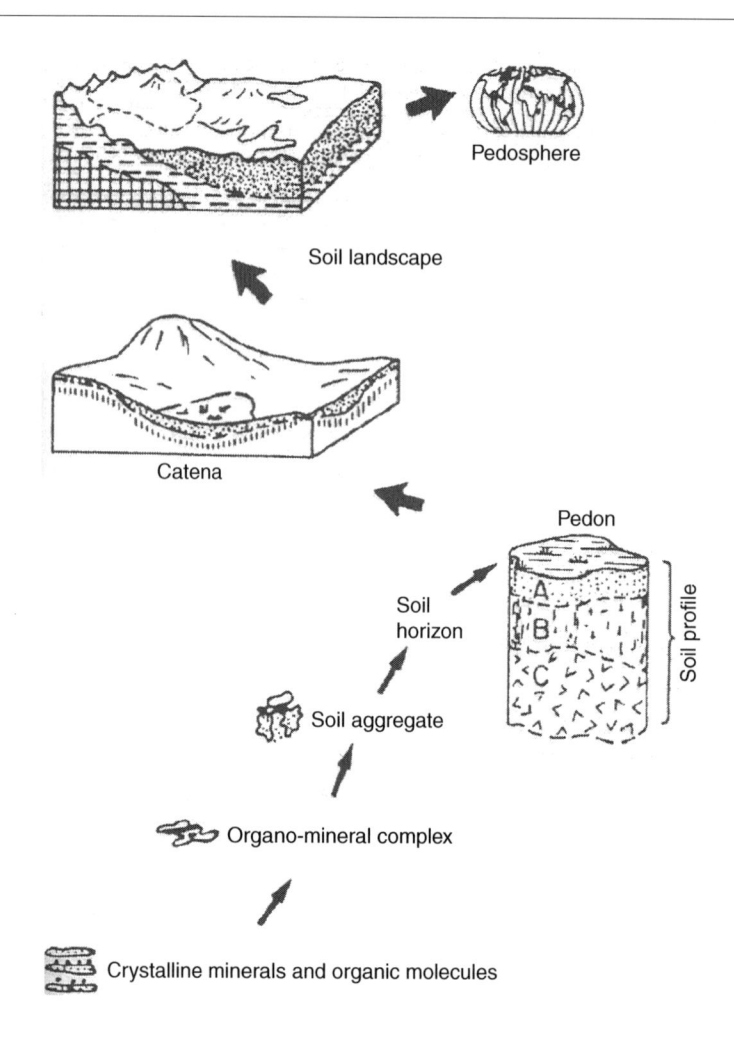

profile is the unit at which we can check upon the outcomes of our management on the soil and choose land for farming or experimental plots. Then, at the level of the *catena* (the linked chain of sites from crest to valley bottom), we make decisions about conservation practices; within the limits of landscapes we establish indices of performance or plan land use – the structure of the sown area and optimum distribution of ploughland, forests and meadows and even the kinds of political and economic organizations. And finally, though rarely considered, the pedosphere is the object of interest and care of mankind according to the principles of the *World Soil Charter* (UN 1983).

Another scenario correlates fluxes in a *soil–crops–atmosphere* system where the key links are humus formation; throughflow and drainage of water; and uptake of nutrients by crops (Fig. 7). This is by no means the complete picture – it does not show the return of carbon to atmosphere as $CO_2$ – but it does emphasize the role of soil as the interface or *membrane of interaction* between the atmosphere and lithosphere. Moreover, far from being an inert membrane, the soil creates local conditions for an immense variety of physical, chemical and biological processes. Whereas the lithosphere, atmosphere and hydrosphere are practically single-phase systems (respectively, solid, gaseous and liquid), the soil embraces all three phases – solid material, air and water; hence it offers richer possibilities. It is ramified and assimilated by every form of life: plant roots, invertebrates, vertebrates, bacteria, fungi, protozoa, viruses – in all their variety of form and function; and it is the scene of interaction of the natural environment and agriculture – in Viliams' words, "from big geological rotations to small biological rotations" (Viliams

**Fig. 7** The soil–crops–atmosphere system (after Sposito and Reginato)

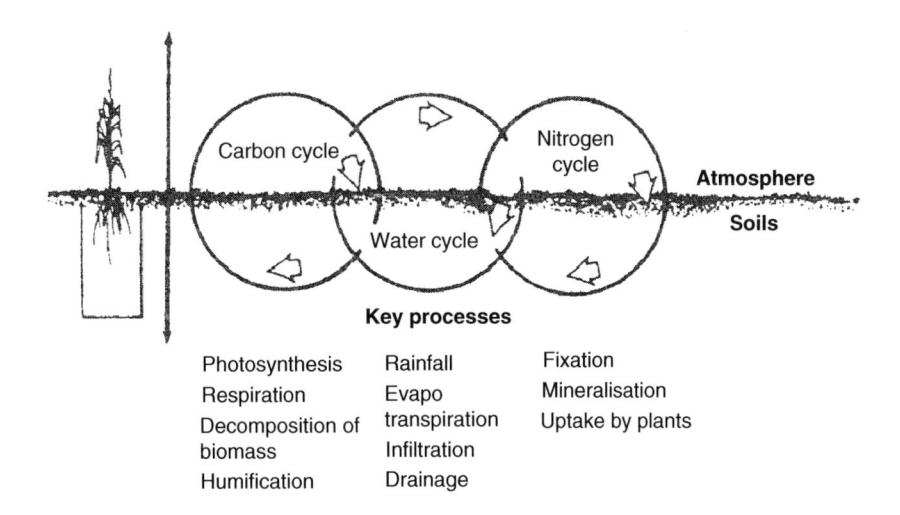

Key processes

Photosynthesis
Respiration
Decomposition of biomass
Humification

Rainfall
Evapotranspiration
Infiltration
Drainage

Fixation
Mineralisation
Uptake by plants

1940). Life on Earth assimilated the interface between the atmosphere and lithosphere and created the soil.

Agriculture makes use of all the components of the biosphere (especially the soil) but, in contrast to the farming systems that we have now, ecological agriculture is biosphere centred. It seeks to maintain the natural biochemical balance of the soil, in particular avoiding losses of humus, nitrogen and of the soil itself that disrupt the biological cycle.

Soil is truly wonderful. Such a soil as the chernozem, which offers especially rich possibilities, should be used wisely and we should learn from the mistakes of the past – striving for maximum yield without appreciating the effect of technology on the soil and the natural ecosystem. Following the pragmatic Americans, our focus should be agriculture on the soil, not on the fields. Indeed, the soil can be enhanced by ecological agriculture which seeks an optimal balance between economic success, ecological functionality and minimization of damage to the soil and the wider environment.

## Soil Sustainability

Sustainability, in the sense of resilience, is something of a catch phrase nowadays – akin to *moral resilience* – that obligatory phrase in the old requirements for going abroad from the Soviet bloc. But soil needs to be resilient – in every dimension but, especially, the vertical (the soil profile) and over time (during the development of an ecosystem or agro-ecosystem).

According to Hitrov (2002): "The sustainability of soil can be regarded as its ability to maintain its state (structure, composition, functionality, space occupied) under slight variations of soil-forming factors; it is also seen as the ability to reconstruct the main qualitative features of its primary state after perturbation."

The first part of this definition is understandable and achievable; self-reconstruction after serious disturbance is more doubtful. The main thing is to preserve the topsoil. The topsoil or A horizon of a full-profile chernozem is more than 40 cm thick (Krupenikov 1978), so the subsoil is not drawn into the plough layer. Under ecological agriculture, the topsoil with all its unique attributes will remain resilient over a long period. However, it is degraded by disturbances such as soil erosion, deep ploughing, or stumping of orchards and vineyards. In the *slightly eroded* category, the thickness of the topsoil may be as little as 17 cm, and the *severely eroded* category has lost all of its topsoil. Continual cultivation weakens the soil's resilience and, once eroded, the soil can never return to its pristine state; so ploughland needs and deserves protection against erosion.

The particle size distribution, mineralogy and the chemical composition of a soil are inherited from the parent material; they are stable under cultivation although they may undergo some changes over a long period of time. But this is not the case with humus – so it is well worthwhile to return plant residues to the soil and practise rotations with perennial crops such as lucerne. Manuring has a special value (Ursu 1986, Turcan 1985) and application of nitrogen and phosphorus fertilizers also slows down the loss of humus in cultivated soils. Viliams, amongst many agronomists, pointed out that the traditional ley grass-arable system is directed to the maintenance of humus and, with it, the architecture of the pore space that determines the soil's water-holding capacity. This issue will be analysed in more detail in the latter part of this book.

Under a leaching climate and under cropping, which carries nutrients off-site, soil is not resistant to the loss of nutrient elements (nitrogen, phosphorus, potassium, sulphur and micro-nutrients). The ecological approach is to accomplish a biologically closed turnover of nutrients to the maximum extent possible. This means that what is harvested with the crop and leached by throughflow must be replaced by manure and fertilizers.

Pollution of soil and water is a different but related issue. As a rule, fertilizing the soil with animal slurry at its normal dilution does not contaminate the soil or water (Andries 2000) but there is a problem of buffering capacity in sandy soils, especially under the excessive application of nitrates that took place in the past (and which can occur nowadays as well). The soil may not suffer from any surplus of nitrates but they are leached through the soil to the groundwater – and the groundwater feeds rivers, lakes and estuaries where excessive and unbalanced amounts of nutrients promote toxic blooms of algae and de-oxygenation of the water which causes fish kills.

The following chapters deal, in turn, with stable and unstable soil attributes that, together, confer resilience to the soil under natural conditions and, to a surprising degree, under cultivation.

## References

Alekseev VE 1999 *Mineralogy of soil formation in steppe and forest-steppe zones of Moldova: diagnosis, parameters, factors, processes.* Stiinta, Chisinau, 240p (Russian)

Andries SV (editor) 2000 *Bulletin for ecopedological monitoring,* 7th edition. Nicolai Dimo Institute of Soil Science and Agrochemistry, Chisinau, 67p (Romanian)

Hitrov NB 2002 The notion of soil resilience to external influence. 3–6 in NB Hitrov (editor) *Soil stability to natural and human influences.* Nauka, Moscow 295p (Russian)

IUSS WRB 2006 *World reference base for soil resources 2006.* IUSS working group WRB. World Soil Resources Rept 103, FAO, Rome

Jenny H 1941 *Factors of soil formation – a system of quantitative pedology.* McGraw-Hill, New York, NY

Krupenikov IA (editor) 1978 *Statistical parameters for soil composition and properties in Moldova. Vol.1.* Stiinta, Chisinau, 180p (Russian)

Krupenikov IA 1979 *Carbonatic chernozem.* Stiinta, Chisinau, 105p (Russian)

Krupenikov IA 1988 Vernadsky-Dokuchaiev biosphere-soil. Celebration of 125 years for VI Vernadsky. *Pochvovedenie* 7, 5–14 (Russian)

Krupenikov IA 2001 *Recommendation for improvement of eroded soils by bringing in soil materials.* Nicolai Dimo Research Institute of Soil Science and Agrochemistry, Chisinau, 50p (Russian)

Krupenikov IA and BP Podimov 1987 *Classification and systematic list of Moldavian soils.* Stiinta, Chisinev, 157p (Russian)

Krupenikov IA, IS Constantinov and GP Dobrovolsky 2002 *Degraded soils and their restoration by using natural ameliorants. A review.* Institute of Economics and Information, Chisinau, 50p (Russian)

Nour DD 2001 *Soil erosion. The essence of the process. Consequence, minimalisation and stabilisation.* Pontos, Chisinau, 427p (Russian)

Sinkevici ZA 1989 *Modern processes in chernozemic soils of Moldova.* Stiinta, Chisinau, 215p (Russian)

Sposito G and Reginato RJ (editors) 1992 *Opportunities in basic soil science research.* Committee on Opportunities in Soil Science, Soil Science Society of America, Madison, WI (Russian translation by Targulian and Gerasimova published by Geos, Moscow 2000)

Turcan MA 1985 *The agrochemical basis for utilization of organic fertilizers.* Stiinta, Chisinau, 275p (Russian)

UN 1983 World Soil Charter. *Pochvovedenie* 7, 7–11

Ursu AT (editor) 1985 *Soils of Moldova. Vol.2 Soil geography, description of soil provinces, districts and sub-districts.* Stiinta, Chisinau, 239p (Russian)

Ursu AT (editor) 1986 *Soils of Moldova. Vol.3 Soil management, protection and improvement.* Stiinta, Chisinau, 333p (Russian)

Vernadsky VI 1965 *The chemical composition of earth's biosphere and its closeness.* Nauka, Moscow, 374p (Russian)

Vernadsky VI 1967 *Biosphere (selected works on biogeochemistry).* Nauka, Moscow, 376p (Russian)

Viliams VR (editor) 1940 *The history of soil fertility. Vol.1. The science of soil fertility in the 19th century.* Selihozgiz, Moscow and Leningrad, 428p (Russian)

# Soil Texture and Structure

**Abstract** The soil's composition and texture provide a physical framework and largely determine its capacity to supply water and nutrients and to develop a soil structure resistant to erosion. The finest fractions, having the largest surface area, are the most active. Soil texture, or particle size distribution, is important in various domains, especially influencing response to tillage. Chernozem inherit a loamy texture from the loess parent material but in situ clay formation is characteristic of *Calcareous* and *Common chernozem*, whereas degradation of clay by weathering occurs in *Podzolized chernozem*. Comparison of the particle size and aggregate size distribution reveals the chernozem's extraordinary capacity for aggregation. According to particle size analysis, silt comprises 60–70% of the soil mass and clay about 15%, but the aggregate size distribution shows almost nothing in these fractions; they are combined into coarser aggregates that lend permeability to air, water and roots, resistance to erosion, and great available water capacity. This soil structure is best developed in pristine chernozem but is remarkably resilient under cultivation.

**Keywords** Soil texture · Soil structure · Aggregation · Weathering · Clay formation

## Soil Texture

The soil's inheritance from its parent material includes its composition and particle size distribution or texture. These provide the physical framework and largely determine the soil's capacity to supply water and nutrients and to develop a soil structure resistant to erosion – which depend on the surface area of the solid particles and the activity of these surfaces. The finest fractions, having the largest surface area, are the most active.

Table 1 shows the texture of the different soil sub-groups in Moldova (Ursu 1985). These textural classes are arbitrary and overlap in the field; the somewhat untidy group of sandy loam-cum-silt loam and clay loam arises because the texture of many Moldovan soils lies close to the boundaries of the standard soil textural classes. Our silt loams and clay loams lie close to the coarse boundaries of these classes and our sandy loams are close to the fine boundary of that class. The important thing is that most of our soils are *loams* – which have a mixture of the finest particles (clay, with an apparent size of <1 μm), middle-sized particles (silt, with a diameter of 1–50 μm) and coarse particles (sand, with a diameter of 50 μm to 1 mm).

The *Typical* and *Leached chernozem* of the north of the country contain somewhat more clay and silt than the southern varieties (Table 2), which is one reason for the greater fertility and *bonitet* rating of the chernozem

**Table 1** Moldavian soils according to soil texture

| | % area of sub-types | | | |
|---|---|---|---|---|
| Soil and sub-types | Sandy loam-cum-silt loam/clay loam | Medium loam | Light loam | Sandy loam and sand |
| *Grey forest* | 43.5 | 30.5 | 18.4 | 7.6 |
| *Dark grey forest* | 65.1 | 22.1 | 10.5 | 2.3 |
| *Chernozem* | | | | |
| *Calcareous* | 60.0 | 35.8 | 3.2 | 1.0 |
| *Common* | 79.0 | 17.8 | 2.2 | 1.0 |
| *Typical* | 89.8 | 8.4 | 0.9 | 0.9 |
| *Leached* | 83.6 | 11.4 | 3.0 | 2.0 |
| *Podzolized* | 76.8 | 20.2 | 2.4 | 0.6 |

**Table 2** Mean silt and clay content in chernozem

| Sub-type | Depth (cm) | n | Clay (%) | Silt (%) |
|---|---|---|---|---|
| *Calcareous* | 0–25 | 208 | 49.9 | 30.2 |
| | 25–50 | 155 | 50.9 | 30.6 |
| | 50–75 | 130 | 49.6 | 28.8 |
| | 75–100 | 108 | 50.9 | 28.6 |
| | 100–150 | 125 | 50.3 | 28.6 |
| | 150–200 | 104 | 48.6 | 27.7 |
| *Common* | 0–25 | 218 | 50.9 | 31.3 |
| | 25–50 | 149 | 46.1 | 28.5 |
| | 50–75 | 139 | 51.9 | 30.6 |
| | 75–100 | 125 | 51.0 | 29.0 |
| | 100–150 | 151 | 51.0 | 28.6 |
| | 150–200 | 160 | 51.0 | 29.4 |
| *Xerophyte-wooded* | 0–25 | 15 | 47.5 | 28.0 |
| | 25–50 | 20 | 49.4 | 29.6 |
| | 50–75 | 21 | 48.4 | 30.0 |
| | 75–100 | 24 | 47.8 | 27.3 |
| | 100–150 | 5 | 48.3 | 25.1 |
| | 150–200 | 8 | 44.3 | 22.4 |
| *Typical* | 0–25 | 80 | 55.7 | 34.0 |
| | 25–50 | 68 | 51.3 | 32.0 |
| | 50–75 | 58 | 56.7 | 34.4 |
| | 75–100 | 43 | 55.2 | 34.5 |
| | 100–150 | 50 | 56.8 | 32.5 |
| | 150–200 | 33 | 54.7 | 32.0 |
| *Leached* | 0–25 | 161 | 51.8 | 33.0 |
| | 25–50 | 103 | 53.6 | 34.1 |
| | 50–75 | 131 | 52.8 | 35.1 |
| | 75–100 | 110 | 53.4 | 33.6 |
| | 100–150 | 113 | 52.8 | 33.9 |
| | 150–200 | 70 | 52.2 | 30.6 |

of northern Moldova. Coarser textures (light loam and sandy loam) have less humus and an acid reaction; they are not so productive for field crops but they are excellent for potatoes and other root crops, and for tobacco, especially Virginia and Berley varieties.

Soil texture is important in various domains, especially influencing response to tillage, so we pay special attention to this property using the careful measurements of a specialist, Dr Eliza Screabina. Going beyond conventional procedures, she also looked at the colloidal fraction (particles with an equivalent diameter of less than 0.0002 mm), and sampled from every 3 cm within the topmost metre of the soil profile, which enables us to make correct comparisons between different soil layers and with the soil parent material (Tables 3 and 4).

The *Typical chernozem* illustrated is from a shelter belt – conditions comparable to arable under ecological agriculture but rather different from the situation under conventional tillage. Its texture straddles the boundary between light clay and clay loam.[1]

From particle size analysis, Krupenikov and Screabina (1976) deduced a process of in situ clay formation in the upper layers of chernozem, especially in *Calcareous chernozem*, which increases the soil's chemical and physical activity.

## Soil Structure

The procedure for particle size analysis is quite severe. The strong reagent breaks down soil aggregates to separate the individual mineral particles; there is no such reagent in nature or in agriculture. To estimate the size distribution of natural soil aggregates, the sample is shaken gently in water – comparable to the action of rainfall or irrigation. Comparison of the particle size and aggregate size distributions lends insight into the soil's real size fractions and, for illustration, we have selected profiles of a *Typical chernozem* (from Rishcani District, similar to the chernozem of the Selectia Experimental Station) and an *Ordinary chernozem*.

For the *Typical chernozem*, the particle size analysis (Table 3) shows no coarse sand (particles greater than 1 mm diameter) and only a little fine-medium sand (0.05–0.1 mm diameter); in contrast, the silt fraction (<0.001 mm) makes up 60–70% of the soil mass and the colloidal fraction 15–30%. Comparing this with the aggregate size distribution (Table 4), we observe a shift of the size fractions: the quantity of micro-aggregates in the 0.1–1 mm category remains greater than one-third of the soil mass to a depth of 70 cm yet, according to the particle size analysis, there is no particulate material of this size. By contrast, silt-sized aggregates comprise less than 1%, and in the parent material only 2 or 3%, although the particle size analysis demonstrates that silt comprises the greater part of the soil mass. The same picture emerges from a comparison of the particle size and aggregate size distribution of the

---

[1] From Screabina's data set we may note that the light clays ($n = 105$) and clay loams ($n = 98$) are close to each other: light clays having on average 63.9% clay and clay loams on average 63.4% clay. Medium clays ($n = 29$) having 66–80% clay and heavy clays ($n = 4$) 81–100% clay are difficult to cultivate, easily compacted and sometimes remained untilled (Krupenikov et al. 1990) but occupy only 0.4% of the country (Ursu 1985).

**Table 3** Particle size distribution of a *Typical chernozem*

| Depth (cm) | % by mass of particle size fractions (mm) | | | | | | | |
|---|---|---|---|---|---|---|---|---|
| | 1.0–0.1 | 0.1–0.05 | 0.05–0.01 | 0.01–0.005 | 0.005–0.001 | <0.001 | <0.0002 | <0.01 |
| 0–3 | 0 | 0 | 42 | 10 | 18 | 30 | 16 | 58 |
| 3–6 | 0 | 0 | 44 | 10 | 20 | 26 | 15 | 56 |
| 6–9 | 0 | 0 | 37 | 9 | 17 | 37 | 22 | 63 |
| 9–12 | 0 | 0 | 38 | 9 | 15 | 38 | 26 | 62 |
| 12–15 | 0 | 0 | 39 | 11 | 15 | 35 | 24 | 61 |
| 15–18 | 0 | 1 | 35 | 9 | 15 | 40 | 31 | 64 |
| 18–21 | 0 | 0 | 39 | 8 | 15 | 38 | 28 | 61 |
| 21–24 | 0 | 0 | 39 | 8 | 17 | 36 | 29 | 61 |
| 24–27 | 0 | 1 | 39 | 6 | 16 | 38 | 30 | 60 |
| 27–30 | 0 | 0 | 40 | 6 | 15 | 39 | 29 | 60 |
| 30–33 | 0 | 0 | 37 | 10 | 14 | 39 | 27 | 63 |
| 33–36 | 0 | 0 | 43 | 6 | 14 | 37 | 31 | 57 |
| 36–39 | 0 | 3 | 38 | 8 | 13 | 38 | 26 | 59 |
| 39–42 | 0 | 2 | 40 | 4 | 15 | 39 | 31 | 58 |
| 42–45 | 0 | 3 | 40 | 4 | 16 | 37 | 29 | 57 |
| 45–48 | 0 | 4 | 39 | 6 | 12 | 39 | 28 | 57 |
| 48–51 | 0 | 4 | 39 | 5 | 9 | 43 | 32 | 57 |
| 51–54 | 0 | 3 | 39 | 9 | 8 | 41 | 34 | 58 |
| 54–57 | 0 | 6 | 36 | 10 | 7 | 41 | 31 | 58 |
| 57–60 | 0 | 7 | 37 | 6 | 11 | 39 | 25 | 56 |
| 60–63 | 0 | 7 | 38 | 5 | 8 | 42 | 32 | 55 |
| 63–66 | 0 | 6 | 38 | 7 | 12 | 37 | 24 | 56 |
| 66–69 | 0 | 6 | 42 | 5 | 7 | 40 | 31 | 52 |
| 69–72 | 0 | 6 | 40 | 5 | 11 | 38 | 27 | 54 |
| 72–75 | 0 | 9 | 35 | 6 | 11 | 39 | 32 | 56 |
| 75–78 | 0 | 10 | 37 | 7 | 9 | 37 | 25 | 53 |
| 78–81 | 0 | 9 | 36 | 6 | 12 | 37 | 23 | 55 |
| 81–84 | 0 | 7 | 39 | 6 | 13 | 35 | 23 | 54 |
| 84–87 | 0 | 3 | 41 | 8 | 12 | 36 | 22 | 56 |
| 87–90 | 0 | 2 | 46 | 4 | 12 | 36 | 25 | 52 |
| 90–93 | 0 | 4 | 44 | 6 | 12 | 34 | 21 | 52 |
| 96–99 | 0 | 2 | 44 | 7 | 16 | 31 | 16 | 54 |
| 100–110 | 0 | 4 | 39 | 6 | 16 | 35 | 18 | 57 |
| 140–150 | 0 | 8 | 34 | 9 | 15 | 34 | 23 | 58 |
| 190–200 | 0 | 8 | 37 | 4 | 16 | 35 | 21 | 55 |

virgin *Common chernozem* (Tables 5 and 6). It contains less silt and clay than the *Typical chernozem* but shows the same remarkable micro-aggregation; the content of free silt is almost zero.

We deduce that the soil and, even, the parent material have a capacity for self-organization into aggregates. This is of the highest importance for the soil's permeability to air and water and its resistance to erosion.

We chose pristine soils to be sure of sampling the topmost layer of the natural soil profile. However, the general picture is similar in ploughland. An old source (Krupenikov 1967) presents textural and micro-structural features of several ploughed soils; these samples were collected from soil genetic horizons – and to great depths to observe buried soils. The example in Table 7, again, shows an absence of free silt, demonstrating the characteristic micro-aggregation of chernozem to a depth of 2 m (which may be observed even at a depth of 4 m).

We may conclude that the chernozem's inheritance from its parent material includes a soil texture favourable to agriculture and a water-stable microstructure that greatly enhances its water-holding capacity, permeability and resistance to erosion. We shall return to each of these attributes.

**Table 4** Micro-aggregate size distribution of a *Typical chernozem*

| Depth (cm) | % mass by aggregate size fraction (mm) | | | | | |
|---|---|---|---|---|---|---|
| | 1.0–0.1 | 0.1–0.05 | 0.05–0.01 | 0.01–0.005 | 0.005–0.001 | <0.001 |
| 0–3 | 32 | 18 | 42 | 6 | 1 | 1 |
| 3–6 | 34 | 19 | 41 | 3 | 2 | 1 |
| 6–9 | 38 | 19 | 34 | 5 | 3 | 1 |
| 9–12 | 30 | 24 | 36 | 6 | 3 | 1 |
| 12–15 | 33 | 17 | 41 | 5 | 3 | 1 |
| 15–18 | 34 | 19 | 40 | 2 | 4 | 1 |
| 18–21 | 36 | 19 | 38 | 2 | 4 | 1 |
| 21–24 | 36 | 19 | 37 | 5 | 2 | 1 |
| 24–27 | 33 | 24 | 35 | 5 | 2 | 1 |
| 27–30 | 42 | 20 | 30 | 4 | 3 | 1 |
| 30–33 | 39 | 13 | 39 | 5 | 3 | 1 |
| 33–36 | 38 | 17 | 34 | 7 | 3 | 1 |
| 36–39 | 37 | 17 | 35 | 7 | 3 | 1 |
| 39–42 | 47 | 23 | 25 | 2 | 2 | 1 |
| 42–45 | 32 | 18 | 41 | 5 | 2 | 2 |
| 45–48 | 32 | 19 | 40 | 6 | 2 | 1 |
| 48–51 | 46 | 18 | 30 | 5 | 2 | 1 |
| 51–54 | 45 | 19 | 28 | 5 | 2 | 1 |
| 54–57 | 31 | 19 | 38 | 8 | 3 | 1 |
| 57–60 | 31 | 20 | 43 | 2 | 3 | 1 |
| 60–63 | 32 | 22 | 36 | 5 | 4 | 1 |
| 63–66 | 31 | 27 | 33 | 5 | 3 | 1 |
| 66–69 | 37 | 18 | 36 | 3 | 4 | 2 |
| 69–72 | 27 | 24 | 39 | 4 | 4 | 2 |
| 78–81 | 28 | 18 | 44 | 2 | 5 | 2 |
| 90–93 | 36 | 19 | 36 | 5 | 2 | 2 |
| 100–110 | 29 | 25 | 37 | 4 | 2 | 3 |
| 140–150 | 22 | 32 | 37 | 4 | 5 | 0 |
| 190–200 | 17 | 32 | 40 | 7 | 4 | 0 |

**Table 5** Particle size distribution of a pristine *Common chernozem*

| Depth (cm) | % by mass particle size fraction (mm) | | | | | | | | |
|---|---|---|---|---|---|---|---|---|---|
| | 1.0–0.25 | 0.25–0.1 | 0.1–0.05 | 0.05–0.01 | 0.01–0.005 | 0.005–0.001 | <0.001 | <0.0002 | <0.01 |
| 0–3 | 13 | 11 | 10 | 31 | 3 | 10 | 22 | 19 | 35 |
| 3–6 | 13 | 10 | 9 | 32 | 2 | 11 | 23 | 20 | 36 |
| 6–9 | 14 | 10 | 9 | 31 | 2 | 10 | 24 | 19 | 36 |
| 9–12 | 14 | 8 | 10 | 33 | 3 | 9 | 23 | 20 | 35 |
| 12–15 | 13 | 9 | 10 | 29 | 6 | 9 | 24 | 20 | 39 |
| 15–18 | 15 | 7 | 13 | 27 | 6 | 8 | 24 | 20 | 38 |
| 18–21 | 15 | 8 | 8 | 31 | 6 | 8 | 24 | 19 | 38 |
| 21–24 | 14 | 9 | 9 | 32 | 4 | 9 | 23 | 19 | 36 |
| 24–27 | 14 | 8 | 9 | 31 | 6 | 9 | 23 | 18 | 38 |
| 27–30 | 14 | 5 | 10 | 33 | 7 | 8 | 23 | 19 | 38 |
| 30–33 | 14 | 10 | 8 | 33 | 5 | 7 | 23 | 16 | 35 |
| 33–36 | 14 | 8 | 9 | 32 | 4 | 10 | 23 | 16 | 37 |
| 36–39 | 13 | 7 | 9 | 34 | 5 | 10 | 22 | 16 | 37 |
| 39–42 | 13 | 7 | 9 | 33 | 6 | 10 | 22 | 16 | 38 |
| 42–45 | 12 | 8 | 10 | 34 | 4 | 10 | 22 | 18 | 36 |
| 45–48 | 14 | 8 | 9 | 33 | 5 | 9 | 22 | 16 | 36 |
| 48–51 | 13 | 8 | 7 | 34 | 6 | 10 | 22 | 16 | 38 |

**Table 5** (continued)

| Depth (cm) | % by mass particle size fraction (mm) | | | | | | | | |
|---|---|---|---|---|---|---|---|---|---|
| | 1.0–0.25 | 0.25–0.1 | 0.1–0.05 | 0.05–0.01 | 0.01–0.005 | 0.005–0.001 | <0.001 | <0.0002 | <0.01 |
| 51–54 | 13 | 7 | 11 | 29 | 5 | 13 | 22 | 16 | 40 |
| 54–57 | 13 | 8 | 10 | 33 | 5 | 10 | 21 | 14 | 36 |
| 57–60 | 12 | 9 | 10 | 31 | 5 | 12 | 21 | 14 | 38 |
| 60–63 | 12 | 8 | 14 | 27 | 8 | 10 | 21 | 13 | 39 |
| 63–66 | 12 | 7 | 20 | 24 | 5 | 12 | 20 | 13 | 37 |
| 66–69 | 12 | 8 | 15 | 27 | 4 | 13 | 21 | 14 | 38 |
| 69–72 | 11 | 6 | 27 | 18 | 6 | 12 | 20 | 13 | 38 |
| 72–75 | 10 | 5 | 29 | 16 | 8 | 13 | 19 | 13 | 40 |
| 75–78 | 9 | 7 | 29 | 18 | 5 | 13 | 19 | 12 | 37 |
| 78–81 | 9 | 4 | 38 | 15 | 3 | 12 | 19 | 11 | 34 |
| 81–84 | 8 | 5 | 30 | 17 | 9 | 12 | 19 | 12 | 40 |
| 84–87 | 9 | 4 | 44 | 10 | 1 | 12 | 20 | 11 | 33 |
| 87–90 | 9 | 6 | 36 | 16 | 2 | 12 | 19 | 11 | 33 |
| 90–93 | 7 | 5 | 44 | 9 | 2 | 15 | 18 | 11 | 35 |
| 93–96 | 8 | 5 | 32 | 16 | 6 | 15 | 18 | 11 | 39 |
| 96–99 | 8 | 5 | 43 | 7 | 7 | 11 | 19 | 12 | 37 |
| 100–110 | 7 | 6 | 31 | 20 | 7 | 11 | 18 | 11 | 36 |
| 140–150 | 6 | 4 | 12 | 43 | 7 | 11 | 17 | 12 | 35 |
| 190–200 | 6 | 6 | 16 | 42 | 6 | 8 | 16 | 12 | 30 |

**Table 6** Micro-aggregate composition of a pristine *Common chernozem*

| Depth (cm) | % by mass aggregate size fraction | | | | | |
|---|---|---|---|---|---|---|
| | 1.0–0.1 | 0.1–0.05 | 0.05–001 | 0.01–0.005 | 0.005–0.001 | <0.001 |
| 0–3 | 55 | 20 | 20 | 3 | 2 | 0 |
| 3–6 | 58 | 18 | 20 | 2 | 1 | 1 |
| 6–9 | 66 | 16 | 13 | 3 | 1 | 1 |
| 9–12 | 70 | 16 | 10 | 2 | 1 | 1 |
| 12–15 | 70 | 14 | 13 | 2 | 1 | 0 |
| 15–18 | 69 | 16 | 18 | 2 | 1 | 0 |
| 18–21 | 60 | 20 | 17 | 2 | 1 | 0 |
| 21–24 | 60 | 19 | 18 | 2 | 1 | 0 |
| 24–27 | 51 | 27 | 18 | 3 | 0 | 1 |
| 27–30 | 53 | 23 | 21 | 2 | 0 | 1 |
| 30–33 | 52 | 24 | 22 | 1 | 1 | 0 |
| 33–36 | 52 | 24 | 21 | 2 | 1 | 0 |
| 36–39 | 53 | 25 | 19 | 2 | 1 | 0 |
| 39–42 | 46 | 25 | 25 | 2 | 2 | 0 |
| 42–45 | 48 | 28 | 20 | 2 | 2 | 0 |
| 45–48 | 47 | 31 | 19 | 2 | 1 | 0 |
| 48–51 | 50 | 25 | 22 | 2 | 1 | 0 |
| 51–54 | 47 | 27 | 23 | 2 | 1 | 0 |
| 54–57 | 45 | 28 | 21 | 3 | 1 | 2 |
| 57–60 | 47 | 31 | 18 | 2 | 1 | 1 |
| 60–63 | 42 | 31 | 22 | 3 | 1 | 1 |
| 63–66 | 43 | 29 | 25 | 1 | 1 | 1 |
| 66–69 | 43 | 32 | 21 | 2 | 1 | 1 |
| 69–72 | 44 | 29 | 21 | 3 | 2 | 1 |
| 78–81 | 34 | 33 | 27 | 2 | 3 | 1 |
| 90–93 | 30 | 35 | 28 | 4 | 2 | 1 |
| 100–110 | 26 | 36 | 30 | 4 | 3 | 1 |
| 140–150 | 19 | 36 | 39 | 2 | 3 | 1 |
| 190–200 | 14 | 35 | 43 | 2 | 4 | 2 |

**Table 7** Comparison of micro-aggregate and particle size distribution of a cultivated *Typical chernozem*

| Depth (cm) | Analysis | Diameter (mm), % by mass | | | | | | | Particles <0.01 mm, % | Structural factor[a] | Aggregates <0.01 mm, % |
|---|---|---|---|---|---|---|---|---|---|---|---|
| | | >0.25 | 0.25–0.1 | 0.1–0.05 | 0.05–0.01 | 0.01–0.005 | 0.05–0.001 | <0.001 | | | |
| 0–20 | Aggregate size | 21 | 9 | 16 | 39 | 7 | 6 | 2 | 68 | 70 | 53 |
| | Particle size | 0 | 0 | 3 | 29 | 16 | 11 | 41 | | | |
| 40–50 | Aggregate size | 23 | 10 | 20 | 33 | 6 | 6 | 2 | 64 | 68 | 50 |
| | Particle size | 0 | 1 | 0 | 35 | 13 | 10 | 41 | | | |
| 90–100 | Aggregate size | 19 | 5 | 21 | 39 | 11 | 5 | 0 | 65 | 75 | 49 |
| | Particle size | 0 | 1 | 0 | 34 | 13 | 9 | 43 | | | |
| 140–150 | Aggregate size | 13 | 13 | 25 | 36 | 8 | 5 | 0 | 62 | 54 | 49 |
| | Particle size | 0 | 0 | 5 | 33 | 9 | 18 | 35 | | | |
| 190–200 | Aggregate size | 10 | 0 | 29 | 41 | 7 | 3 | 0 | 61 | 70 | 51 |
| | Particle size | 0 | 1 | 2 | 36 | 9 | 11 | 41 | | | |
| 290–300 | Aggregate size | 5 | 7 | 28 | 50 | 7 | 3 | 0 | 63 | 64 | 53 |
| | Particle size | 0 | 0 | 3 | 34 | 10 | 14 | 39 | | | |
| 400–410 | Aggregate size | 4 | 9 | 27 | 50 | 7 | 3 | 0 | 52 | 37 | 42 |
| | Particle size | 0 | 0 | 5 | 43 | 9 | 16 | 27 | | | |

[a]Structural factor according to Kachinshi (Atamaniuc and others 1970, Gradusov 1990).

# References

Atamaniuc AK, RM Vladimir and LS Karapetean 1970 *Physical and ameliorative properties of soils in Moldova*. Stiinta, Chisinau, 70p (Russian)

Gradusov BP 1990 Petrographic-mineralogical composition and properties important for soil fertility and environmental protection. 3–6 in *Mineralogical composition and microstructure of soils*. Russian Academy of Agricultural Sciences, Moscow (Russian)

Krupenikov IA 1967 *Chernozem of Moldova*. Cartea Moldoveneasca, Chisinau, 427p (Russian)

Krupenikov IA and IE Screabina 1976 Clayization of chernozem from the Danube region. *Pochvovedenie* 11, 3–13 (Russian)

Krupenikov IA, VE Alexeev, BP Podimov, IE Screabina and BHI Smirnov 1990 *Clayed soils of Moldova (Genesis, Properties, Evolution, Arable Use)*. Stiinta, Chisinau, 167p (Russian)

Ursu AT (editor) 1985 *Soils of Moldova. Vol.2 Soil geography, description of soil provinces, districts and sub-districts*. Stiinta, Chisinau, 239p (Russian)

# Soil Mineralogy and Elemental Composition

**Abstract** Mineralogical analysis tells us what the soil is made of and charts the course of soil evolution through the processes of clay formation, accumulation, translocation and the destruction of minerals by weathering. We may consider the evolution of the varieties of chernozem, from the *Calcareous* to the *Podzolized*, as a sequence of weathering and leaching – although this is a very slow process from the human standpoint, taking place over millennia. Chernozem are rich in felspar, mica and illite. In the case of felspar, there is no difference in content between the soil and the parent material, which indicates that weathering is weak and that the soil holds valuable mineral reserves. In the case of mica, there is less in the soil compared with the parent material but a commensurate increase in the clay mineral illite, a weathering product of mica. This is significant for plant nutrition: the major plant nutrient potassium is an important component of potassium felspar, mica and illite. Weathering of these minerals releases potassium into the soil, which accounts for the generally low effectiveness of potassic fertilizers in chernozem; the natural reserves will suffice for centuries. Another perspective may be gleaned from the soil's elemental composition. For all their differences in detail, the composition of all varieties of chernozem is much the same – and unique in the world of soils; there is no lack of any nutrient element, which is the most important feature for agriculture. The major elements are conventionally expressed as oxides; on this basis, 60–70% is silica ($SiO_2$) and 13–16% alumina ($Al_2O_3$). There are also substantial amounts of iron (5–6%) and silicate calcium (about 2%); translated into a metre thickness of soil these come to hundreds of tonnes per hectare. However, there is not so much phosphorus, which is the nutrient in shortest supply.

**Keywords** Soil mineralogy · Weathering · Soil evolution · Plant nutrients · Soil fertility · Total chemical analysis

## Mineralogy

Particle size analysis tells us about the soil's mechanical composition; mineralogical analysis tells us what the soil is made of in terms of individual minerals and, which may be equally important for agriculture, what it lacks. Mineralogy also charts the course of soil evolution and provides landmarks as individual minerals change in the course of weathering.

In one sense, mineralogy was the mother of soil science or, at least, the midwife: Dokuchaev was professor of mineralogy at the University of St Petersburg from 1879 to 1896, teaching in one of the first buildings of the northern capital, on Vasilev's Island. One of the present authors had the privilege of visiting the historic house and sitting at the desk where Dokuchaev wrote *Russian chernozem, our steppe: its past and present*, and many volumes on the soils of Nizhniy Novgorod and Poltava provinces. N.M. Sibirtsev, A. Ferhmin, K.D. Glinka and L.L. Prasolov later occupied the same chair, and V.I. Vernadsky was a graduate of the same department.

It is one of the paradoxes of soil science that mineralogy was subsequently neglected. But more recently, in Moldova, V.E. Alekseev created a well-equipped laboratory and worked energetically in this field, publishing the *Mineralogy of soil formation in the steppe and forest steppe of Moldova* (Alekseev 1999). His novel methods of simultaneous quantitative analysis of primary minerals (quartz, felspar, mica, etc.)

and clay minerals (smectite, kandite, etc.) enable us to calculate mineralogical ratios and trends of different soils. In this way, we may follow the path of soil evolution – the processes of clay formation, accumulation, translocation without chemical destruction, the destruction of minerals and their leaching in solution.

From Alekseev's data, we abstract a few details to illustrate the characteristics of the chernozem. Its mineral composition depends on the parent material. The so-called *primary minerals*, quartz and a few minerals with a more complex composition, are derived from the native rock. These are transformed by weathering into *secondary minerals*, in particular clay minerals which are very reactive, both physically and chemically, owing to their enormous surface area and capacity to absorb both water and nutrients. We can see these effects if we compare fine-textured soils (heavy loams) and coarse-textured soils (light loams), even within the chernozem family.

Table 1 shows the mineralogical composition of a heavy loam chernozem. Considering the content of felspar, there is no difference between the soil and the parent material (both contain a little under 16%), which indicates that weathering is weak and that the soil holds valuable mineral reserves. In the case of mica, there is 9% in the parent material and less than 6% in the soil; the clay mineral illite is a weathering product of mica and we see nearly 13% in the soil compared with a little over 8% in the parent material. These differences are statistically significant at 95% probability – and significant for plant nutrition: the major plant nutrient potassium is an important component of potassium felspar, mica and illite, and weathering of these minerals releases potassium into the soil. The very active clay mineral smectite is somewhat less common in soil than in the parent material but, with a content of 15–19%, there is plenty.

If we compare the mineralogical composition of heavy loam chernozem with clay chernozem (Tables 1 and 2, respectively), we see that the clays contain a comparable amount of felspar but much less mica and more smectite and illite; again, the differences are statistically significant. There is no reason to think that the clayey chernozem are older and, therefore, more strongly weathered than the loamy so we conclude that mineralogy, although changeable in the course of soil formation, is largely inherited from the parent material.

**Table 1** Mineral composition of heavy loam chernozem, $n = 22$

| Soil 1 Rock 2 | Mean (%) | Std deviation | CoV (%) |
|---|---|---|---|
| Felspar | | | |
| 1 | 15.4 | 1.6 | 10 |
| 2 | 15.8 | 2.3 | 15 |
| Mica | | | |
| 1 | 5.7 | 1.5 | 26 |
| 2 | 9.0 | 2.2 | 25 |
| Smectite | | | |
| 1 | 15.9 | 2.0 | 13 |
| 2 | 18.8 | 4.3 | 23 |
| Illite | | | |
| 1 | 12.9 | 2.9 | 23 |
| 2 | 8.2 | 2.3 | 28 |

**Table 2** Mineralogical composition of clay chernozem, $n = 5$

| Soil 1 Rock 2 | Mean (%) | Std deviation | CoV (%) |
|---|---|---|---|
| Felspar | | | |
| 1 | 10.0 | 2.6 | 26 |
| 2 | 9.3 | 3.7 | 39 |
| Mica | | | |
| 1 | 6.7 | 3.0 | 45 |
| 2 | 9.1 | 4.3 | 48 |
| Smectite | | | |
| 1 | 27.4 | 8.6 | 31 |
| 2 | 32.9 | 10.2 | 30 |
| Illite | | | |
| 1 | 14.3 | 3.8 | 26 |
| 2 | 11.0 | 2.0 | 18 |

Alekseev is a rich vein of information. Comparing the distribution of silt and clay in the chernozem with the parent material, he points out that the clay fraction increases only in the early stages of soil formation, especially in the *Calcareous chernozem* (Screabina 1972, Krupenikov and Screabina 1976). In the *Typical chernozem* and *Podzolized chernozem*, there is even some degradation of clay. We may consider the evolution of the varieties of chernozem, from the *Calcareous* to the *Podzolized*, as a sequence of weathering and leaching (Krupenikov 1967) although this is a very slow process from the human standpoint, taking place over millennia.

Several primary and clay minerals contribute to the chernozem's fertility, in particular illite which is a rich source of available potassium and which accounts for the generally low effectiveness of potassic fertilizers in chernozem; the natural reserves will suffice for centuries. Agriculture, especially irrigation, accelerates the processes of weathering (Pozneac 1997, Prihodiko 1996, Krupenikov 1995) but, in the case of the chernozem, the mineralogical inheritance will last for a long time – and a good thing too, for it is the foundation of the soil's mechanical and chemical resilience, and fertility. B. Gradusov considers it to be the foundation of the world's ecosystems; nearly all of the most densely populated parts of the world are characterized by a rich mineralogical inheritance (Gradusov 1990, 1995), and so it is in Moldova. We should take this inheritance into account in rebuilding of our agriculture and agricultural landscape.

## Total Chemical Analysis

Another perspective on the fundamental attributes of the soil may be gleaned from its elemental composition. Total chemical analysis is a difficult and expensive procedure, not undertaken nowadays in Moldova. However, since elemental composition is very stable, we may make use of old analyses carried out by V. Ganenko, G. Strijova and L. Biriukov at the Institute of Pedology and Agrochemistry, under Dimo. Table 3 presents data for *Typical chernozem*, similar to the soil at Selectia Experimental Station. The elements are conventionally expressed as oxides; on this basis, 60–70% is silica ($SiO_2$) and 13–16% alumina ($Al_2O_3$). There are also substantial amounts of iron (5–6%) and silicate calcium (about 2%); translated into a metre thickness of soil comes to hundreds of tonnes per hectare. However, there is not so much phosphorus, which is the nutrient in shortest supply.

Table 4 illustrates similarities and differences between chernozem sub-types from different parts of the country.

The *Common chernozem* is close to the *Typical* in terms of silica, alumina and iron: cf. Tables 3 and 4, profiles 272 (silty clay) and 506 (sandy loam); the calcium stock is especially high; and both soil and parent material are rich in sulphur. The low content of phosphorus in the upper part of profile 272 might be explained by long-term arable use that strips phosphorus from the soil by carrying away the crops.

The *Leached chernozem* (profile 400) shows a different elemental profile. Leaching is reflected in the downward migration of alumina and iron; sulphur decreases in the top layers, calcium in the subsoil (although calcium reserves remain large).

Chernozem from the south of the country are represented by *Xerophyte-wooded chernozem* (profiles 231 and 238) and a *Calcareous chernozem* (profile 241).

For all their differences in detail, the elemental composition of all chernozem is much the same – and unique in the world of soils; there is no lack of any nutrient element, which is the most important feature for agriculture. Strijova (cf. Ursu 1984) reviewed the total chemical composition of the top 50 cm of chernozem from the whole country (Table 5). There are only a few samples from some types, which precludes statistical analysis, but the results are germane – in particular the data for the silt fraction which, through weathering, provides elements available to plants. Inert quartz comprises half to three-quarters of the silt fraction but the alumina and iron components are also substantial (13–19 and 4–9%, respectively) and

**Table 3** Total elemental composition of light clay *Typical chernozem*, % of ignited, carbonate-free material

| Profile | Depth (cm) | $SiO_2$ | $TiO_2$ | $Al_2O_3$ | $Fe_2O_3$ | MnO | CaO | MgO | $Na_2O$ | $K_2O$ | $P_2O_5$ | $SO_3$ | $SiO_2:Al_2O_3$ | $SiO_2:Fe_2O_3$ | $SiO_2:R_2O_3$ |
|---|---|---|---|---|---|---|---|---|---|---|---|---|---|---|---|
| 21 | 0–20 | 72.2 | 0.7 | 14.5 | 5.3 | 0.3 | 2.0 | 1.8 | 0.9 | 2.1 | 0.1 | 0.4 | 8.5 | 36.5 | 6.9 |
|  | 40–50 | 72.1 | 0.6 | 13.4 | 5.2 | 0.4 | 2.2 | 2.6 | 0.9 | 2.1 | 0.1 | 0.5 | 9.1 | 36.5 | 7.3 |
|  | 90–100 | 72.7 | 0.6 | 14.5 | 4.7 | 0.3 | 1.7 | 1.7 | 1.0 | 2.0 | 0.1 | 0.2 | 8.6 | 42.0 | 7.1 |
|  | 140–150 | 72.6 | 0.7 | 13.1 | 5.6 | 0.3 | 1.9 | 1.8 | 1.1 | 1.9 | 0.1 | 0.3 | 9.4 | 34.6 | 7.4 |
|  | 190–200 | 73.7 | 0.7 | 13.9 | 6.1 | 0.3 | 1.9 | 1.8 | 0.9 | 1.8 | 0.1 | 0.3 | 9.0 | 32.4 | 7.0 |
| 27 | 0–20 | 72.8 | 0.6 | 13.2 | 4.8 | 0.1 | 2.1 | 2.7 | 1.7 | 2.8 | 0.1 | 0.5 | 9.3 | 40.4 | 7.6 |
|  | 40–50 | 75.0 | 0.7 | 13.1 | 4.8 | 0.1 | 1.2 | 1.3 | 1.6 | 2.3 | 0.1 | 0.3 | 9.6 | 43.0 | 7.8 |
|  | 90–100 | 69.3 | 0.7 | 16.0 | 4.8 | 0.1 | 3.1 | 1.3 | 1.9 | 2.1 | 0.1 | 0.4 | 7.2 | 38.0 | 6.0 |
|  | 190–200 | 68.2 | 0.7 | 18.8 | 5.0 | 0.1 | 2.7 | 1.6 | 1.7 | 2.0 | 0.1 | 0.4 | 6.3 | 38.0 | 5.4 |
|  | 340–350 | 67.4 | 0.6 | 18.5 | 4.1 | 0.1 | 2.7 | 1.5 | 1.8 | 2.1 | 0.1 | 0.4 | 6.2 | 45.0 | 5.5 |
|  | 490–500 | 65.8 | 0.6 | 22.3 | 6.0 | 0.1 | 0.3 | 2.9 | 2.07 | 3.1 | 0.1 | 0.4 | 5.0 | 30.0 | 4.3 |

**Table 4** Total elemental composition of Moldavian chernozem, % ignited, carbonate-free material

| Profile | Depth (cm) | $SiO_2$ | $Al_2O_3$ | $Fe_2O_3$ | MnO | CaO | MgO | $Na_2O$ | $K_2O$ | $P_2O_5$ | $SO_3$ | Oxide ratios $SiO_2:Al_2O_3$ | $SiO_2:Fe_2O_3$ | $SiO_2:R_2O_3$ |
|---|---|---|---|---|---|---|---|---|---|---|---|---|---|---|
| 272 | 0–29 | 73.9 | 14.6 | 4.1 | 0.2 | 1.4 | 0.8 | – | 2.8 | 0.1 | 0.3 | 8.6 | 47.0 | 7.3 |
| | 40–50 | 73.8 | 14.6 | 4.9 | 0.2 | 1.6 | 0.9 | 1.9 | 2.4 | 0.1 | 0.6 | 8.6 | 40.0 | 7.1 |
| | 70–80 | 73.2 | 14.9 | 4.9 | 0.1 | 2.1 | 1.1 | 1.8 | 2.1 | 0.2 | 0.5 | 8.4 | 39.0 | 7.0 |
| | 90–100 | 74.0 | 14.5 | 4.9 | 0.1 | 1.3 | 1.2 | 2.0 | 2.1 | 0.1 | 0.4 | 8.7 | 40.0 | 7.1 |
| | 190–200 | 74.1 | 14.2 | 4.6 | 0.1 | _ | _ | 2.1 | 2.1 | 0.1 | 0.4 | 8.8 | 42.0 | 7.3 |
| | 330–340 | 74.3 | 14.8 | 4.7 | 0.1 | 1.7 | 0.9 | 1.0 | 2.2 | 0.1 | 0.5 | 8.5 | 43.0 | 7.1 |
| 506 | 0–10 | 71.6 | 14.6 | 5.6 | 0.1 | 2.3 | 1.2 | 1.5 | 2.2 | 0.3 | 0.3 | 8.4 | 33.9 | 6.8 |
| | 50–60 | 70.9 | 16.0 | 4.0 | 0.2 | 2.6 | 1.7 | 1.8 | 2.0 | 0.3 | 0.4 | 7.7 | 46.8 | 6.6 |
| | 90–100 | 71.0 | 14.5 | 5.5 | 0.2 | 2.4 | 1.8 | 1.4 | 1.7 | 0.3 | 0.5 | 8.6 | 34.2 | 6.8 |
| | 140–150 | 71.1 | 14.3 | 5.1 | 0.1 | 2.3 | 1.7 | 2.0 | 1.9 | 0.2 | 0.3 | 8.6 | 37.1 | 7.0 |
| | 190–200 | 72.5 | 15.6 | 5.0 | 0.2 | 1.6 | 1.9 | 1.3 | 1.6 | 0.2 | 0.3 | 8.1 | 38.7 | 6.7 |
| | 280–290 | 73.2 | 16.6 | 3.5 | 0.1 | 2.0 | 1.8 | 0.8 | 1.5 | 0.2 | 0.3 | 7.6 | 56.0 | 6.7 |
| | 340–350 | 73.1 | 14.3 | 4.9 | 0.2 | 2.9 | 1.5 | 0.9 | 1.6 | 0.2 | 0.3 | 8.9 | 39.6 | 7.3 |
| | 390–400 | 73.2 | 15.5 | 4.4 | 0.2 | 1.7 | 2.1 | 0.3 | 1.5 | 0.2 | 0.3 | 8.2 | 44.7 | 6.9 |
| | 460–470 | 74.4 | 14.5 | 4.7 | – | 1.0 | 2.1 | 1.1 | 1.4 | 0.3 | 0.1 | 8.9 | 42.0 | 7.4 |
| 400 | 0–10 | 78.5 | 11.8 | 2.8 | 0.3 | 2.8 | 1.7 | 2.0 | 2.2 | 0.2 | 0.2 | 11.3 | 75.0 | 9.9 |
| | 30–40 | 77.5 | 12.0 | 2.9 | 0.3 | 2.2 | 1.6 | 1.9 | 1.8 | 0.1 | 0.1 | 10.9 | 71.3 | 9.4 |
| | 50–60 | 77.1 | 13.5 | 3 2 | 0.2 | 1.3 | 1.4 | 2.3 | 3.0 | 0.1 | 0.2 | 9.7 | 63.5 | 8.4 |
| | 70–80 | 74.4 | 13.5 | 3.4 | 0.2 | 1.0 | 1.3 | 9.6 | 9.4 | 0.1 | 0.3 | 9.4 | 58.5 | 8. 1 |
| | 100–110 | 77.2 | 14.0 | 3.8 | 0.3 | 1.0 | 1.2 | 2.1 | 2.2 | 0.1 | 0.2 | 9.4 | 54.5 | 8.0 |
| | 180–190 | 77.4 | 12.1 | 4.3 | 0.3 | 0.7 | 1.3 | 1.1 | 2.3 | 0.2 | 0.1 | 10.8 | 50.3 | 8.8 |
| | 290–300 | 79.5 | 11.0 | 2.6 | 0.2 | 1.3 | 1.5 | 2.0 | 1.4 | 0.2 | 0.2 | 12.4 | 51.0 | 9.9 |
| | 400–410 | 80.1 | 10.1 | 3.2 | 0.2 | 0.9 | 1.5 | 1.5 | 1.7 | 0.2 | 0.2 | 13.3 | 58.0 | 10.8 |
| | 460–470 | 83.1 | 8.7 | 2.4 | 0.3 | 0.4 | 1.3 | 1.8 | 1.6 | 0.1 | 0.1 | 16.3 | 91.6 | 13.8 |
| 231 | 0–10 | 76.6 | 13.1 | 3.9 | 0.1 | 1.7 | 1.6 | 0.7 | 1.8 | 0.2 | 0.4 | 9.8 | 53.4 | 8.3 |
| | 30–40 | 75.7 | 13.0 | 4.2 | 0.1 | 1.6 | 1.5 | 1.2 | 1.7 | 0.2 | 0.4 | 9.7 | 48.5 | 8.1 |
| | 50–60 | 77.3 | 13.7 | 4.2 | 0.1 | 1.7 | 1.5 | 0.9 | 1.7 | 0.2 | 0.3 | 9.9 | 49.7 | 8.3 |
| | 90–100 | 74.7 | 12.4 | 4.8 | 0.1 | 1.9 | 2.0 | 0.9 | 1.6 | 0.2 | 0.2 | 10.3 | 41.4 | 8.3 |
| | 190–200 | 76.4 | 12.5 | 4.7 | 0.1 | 1.1 | 2.3 | 0.7 | 1.3 | 0.2 | 0.3 | 10.6 | 43.8 | 8.5 |
| | 340–350 | 77.5 | 12.6 | 4.1 | 0.1 | 1.1 | 2.3 | 0.8 | 1.8 | 0.2 | 0.2 | 10.7 | 51.6 | 8.9 |
| 238 | 0–10 | 72.9 | 12.6 | 5.0 | 0.1 | 2.0 | 1.6 | 2.7 | 2.8 | 0.2 | – | 10.1 | 39.2 | 8.9 |
| | 60–70 | 73.3 | 14.3 | 4.9 | 0.1 | 1.5 | 1.6 | 2.5 | 2.4 | 0.2 | – | 9.1 | 45.3 | 8.8 |
| | 90–100 | 73.8 | 13.8 | 4.7 | 0.1 | 1.6 | 1.4 | 2.6 | 2.3 | 0.2 | – | 9.4 | 42.5 | 8.8 |
| | 190–200 | 72.6 | 13.2 | 4.8 | 0.1 | 3.0 | 2.0 | 2.3 | 2.0 | 0.2 | – | 9.7 | 40.4 | 8.9 |
| | 230–240 | 72.7 | 12.2 | 4.9 | 0.1 | 4.5 | 2.0 | 2.0 | 1.8 | 0.2 | – | 10.7 | 39.1 | 8.9 |
| 241 | 0–10 | 75 5 | 13.1 | 4.5 | 0.1 | 1.5 | 1.8 | 1.5 | 1.7 | 0.2 | 0.3 | 9.7 | 44.6 | 8.0 |
| | 30–40 | 74.9 | 13.6 | 4.6 | 0.1 | 1.9 | 2.1 | 1.4 | 1.9 | 0.9 | 0.3 | 9 3 | 43.0 | 7.7 |
| | 50–60 | 73.7 | 13.2 | 4.7 | 0.1 | 1.0 | 2.3 | 1.3 | 1.4 | 0.4 | 0.3 | 9.4 | 42.4 | 7.7 |
| | 120–130 | 71.8 | 13.6 | 4.7 | 0.1 | 3.2 | 3.0 | 1.2 | 1.3 | 0.2 | 0.3 | 8.9 | 41.1 | 7.3 |
| | 190–200 | 72.4 | 13.2 | 4.7 | 0.1 | 1.0 | 3.7 | 1.4 | 1.8 | 0.2 | 0.5 | 9.3 | 41.0 | 7.6 |
| | 290–300 | 73.1 | 13.1 | 4.6 | 0.1 | 1.3 | 3.6 | 1.5 | 1.7 | 0.1 | 0.3 | 9.5 | 41.8 | 7.8 |
| | 390–400 | 74.8 | 14.0 | 4.5 | 0.1 | 1.0 | 3.1 | 1.7 | 1.6 | 0.2 | 0.4 | 9.2 | 44.2 | 7.6 |
| | 460–470 | 75.3 | 11.1 | 3.6 | 0.1 | 2.6 | 3.1 | 1.7 | 1.5 | 0.2 | 0.3 | 11.4 | 56.2 | 9.5 |

silicate minerals provide two-thirds of the calcium in silt fraction.

Particle size distribution, mineralogy and elemental composition are related to one another although each is studied according to quite different methods. They determine much of the soil's development and productivity within both natural and agricultural ecosystems. In the case of chernozem everywhere, they provide a very good foundation for farming.

# Total Chemical Analysis

**Table 5** Elemental composition of chernozem and silt, % ignited soil 0–50 (cm)

| | $SiO_2$ | | | | $Al_2O_3$ | | | | $Fe_2O_3$ | | | | $Na_2O$ | | | | $K_2O$ | | | |
|---|---|---|---|---|---|---|---|---|---|---|---|---|---|---|---|---|---|---|---|---|
| | 1 | 2 | 3 | 4 | 1 | 2 | 3 | 4 | 1 | 2 | 3 | 4 | 1 | 2 | 3 | 4 | 1 | 2 | 3 | 4 |
| *Chernozem* | | | | | | | | | | | | | | | | | | | | |
| Mean | 68.7 | 70.6 | 79.3 | 74.3 | 11.2 | 11.9 | 11.6 | 13.2 | 3.9 | 3.7 | 3.9 | 4.4 | 1.5 | 0.9 | 0.5 | 1.2 | 1.4 | 1.7 | 1.7 | 2.0 |
| $S_x$ | 0.4 | 0.5 | – | 0.4 | 0.2 | 0.5 | – | 0.3 | 0.1 | 0.2 | – | 0.2 | 0.11 | 0.03 | | 0.2 | 0.04 | 0.13 | – | 0.1 |
| $S$ | 1.0 | 1.2 | – | 1.5 | 0.3 | 1.3 | – | 0.9 | 0.1 | 0.5 | – | 0.7 | 0.22 | 0.07 | | 0.6 | 0.08 | 0.3 | – | 0.5 |
| CoV % | 1.4 | 1.7 | – | 2.0 | 2.6 | 11.1 | – | 6.7 | 2.8 | 13.6 | – | 15.3 | 15.1 | 7.8 | | 48.9 | 5.6 | 18.2 | – | 25.0 |
| $n$ | 4 | 6 | 2 | 14 | 4 | 6 | 2 | 12 | 4 | 6 | 2 | 13 | 4 | 6 | 2 | 15 | 4 | 6 | 2 | 15 |
| *Silt* | | | | | | | | | | | | | | | | | | | | |
| Mean | 46.1 | 43.7 | 46.6 | – | 19.7 | 1.7 | 19.4 | – | 8.9 | 9.7 | 9.2 | – | 0.43 | 0.61 | – | – | 2.1 | 2.7 | – | – |
| $S_x$ | 0.7 | 0.5 | – | – | 0.07 | 0.2 | – | – | 0.5 | 0.2 | – | – | 0.04 | 0.07 | – | – | 0.04 | 0.2 | – | – |
| $S$ | 1.3 | 1.3 | – | – | 1.3 | 5.3 | – | – | 0.9 | 0.5 | – | – | 0.08 | 0.18 | – | – | 0.1 | 0.4 | – | – |
| CoV % | 2.8 | 3.0 | – | – | 6.8 | 33.9 | – | – | 10.4 | 13.6 | – | – | 18.7 | 3.2 | – | – | 4.5 | 13.4 | – | – |
| $n$ | 4 | 6 | 2 | – | 4 | 6 | 2 | – | 4 | 6 | 2 | – | 4 | 6 | – | – | 4 | 6 | – | – |

1. Southern chernozem, 2. Central chernozem, 3. Northern chernozem, 4. Median values for Moldavian chernozem.

# References

Alekseev VE 1999 *Mineralogy of soil formation in steppe and forest-steppe zones of Moldova: diagnosis, parameters, factors, processes*. Stiinta, Chisinau, 240p (Russian)

Gradusov BP 1990 Petrographic-mineralogical composition and properties important for soil fertility and environmental protection. 3–6 in *Mineralogical composition and microstructure of soils*. Lenin All-Soviet Academy of Agricultural Sciences, Moscow (Russian)

Gradusov BP 1995 On evaluating the composition and properties of the lithosphere as a basis for world ecosystems. *Pochvovedenie* 2, 217–225 (Russian)

Krupenikov IA 1967 *Chernozem of Moldova*. Cartea Moldoveneasca, Chisinau, 427p (Russian)

Krupenikov IA 1995 Irrigation of chernozem from the Danube-Pontic region. *Pochvovedenie* 1, 122–127 (Russian)

Krupenikov IA and IE Screabina 1976 Clayization of chernozem from the Danube region. *Pochvovedenie* 11, 3–13 (Russian)

Pozneac SP 1997 *Irrigated chernozem from the south-western part of Ukraine*. VITA, L'viv, 240p (Russian)

Prihodiko VE 1996 *Irrigated steppe soils: functionality, ecology, productivity*. INTELECT, Moscow, 168p (Russian)

Screabina IE 1972 *The process of clayization in chernozem soils of Moldova*. (short version of dissertation) MV Lomonosov State University, Moscow, 28p (Russian)

Ursu AT (editor) 1984 *Soils of Moldova. Vol.1 Genesis, ecology, classification and systematic description of soils*. Stiinta, Chisinau, 351p (Russian)

# Adsorption Capacity and Reaction

**Abstract** Soil fertility depends on the soil's capacity to adsorb bases such as potassium and calcium, retain them against leaching and release them to plants. Adsorption capacity depends on surface-active mineral and organic colloids. The clay mineralogy and colloidal humus of the chernozem provide a high adsorptive capacity and, also, buffer capacity – the ability to resist degradation processes. The adsorbing complex is near-saturated with calcium and magnesium that promote micro-aggregation and resist both acidification and degradation by accumulation of exchangeable sodium. Most arable crops prefer a neutral to slightly acid reaction (pH 7.5–6.0) and the reaction of chernozem topsoils lies within these limits.

**Keywords** Base exchange · Adsorption capacity · Colloidal humus and clay · Exchangeable ions · Reaction · Weathering

## Adsorption Capacity

Soil fertility depends on the soil's capacity to adsorb bases, such as ammonia, potassium and calcium, retain them against leaching but release them to plants. This phenomenon was first investigated in the middle of the nineteenth century by Thomson and Way, in England. Later, fundamental work on adsorption was carried out by our fellow-countryman from Bendery, Konstantin Gedroits, who introduced the terms adsorptive capacity, adsorptive energy and adsorbed (or exchangeable) ions. In this context we should note that as well as basic ions, or cations, some anions such as sulphate are also adsorbed.

The process of ion exchange depends on surface-active mineral and organic colloids. In the chernozem, its colloidal humus and its clay mineralogy (rich in smectite and illite) provide a high adsorptive capacity. They also provide buffer capacity which is the ability to resist degradation processes. The adsorbing complex is near-saturated with calcium and magnesium that resist acidification and, also, degradation by accumulation of exchangeable sodium. These same adsorbed bases also promote micro-aggregation. Table 1 presents summary data on exchangeable calcium and magnesium for 1628 chernozem soil profiles to a depth of 1 m.

The chernozem's ion exchange complex is ideal for agriculture in terms of both the amounts of exchangeable bases and their ratios one to another. The sum of exchangeable cations (in soils lacking exchangeable acidity, this is equivalent to the adsorption capacity) gradually decreases with depth – in concert with the humus content.

AD Fokin's maxim: "Humus is the guardian of soil fertility" (Fokin 1975) might equally well be said of calcium. It can ameliorate the *solonetz* – calcium from gypsum can replace sodium on the adsorbing complex; and the *podzol* – calcium from lime can replace exchangeable acidity. The behaviour of adsorbed magnesium, however, is ambiguous; in small amounts it behaves like calcium but, in large amounts, it behaves rather like sodium – for instance, in rare *magnesium solonetz*. In chernozem generally, the ratio of calcium to magnesium is 5–9:1.

**Table 1** Exchangeable bases in chernozem (mg/100 g soil – determination of exchangeable sodium is complicated by the presence of soluble sodium salts which must be leached before analysis of exchangeable sodium)

| Chernozem sub-types | Ca | | Mg | | Ca:Mg |
|---|---|---|---|---|---|
| | Mean | CoV % | Mean | CoV % | |
| *0–20 cm* | | | | | |
| *Calcareous* | 27.9 | 21 | 3.0 | 43 | 9:1 |
| *Common* | 31.3 | 17 | 3.6 | 40 | 8:1 |
| *Typical* | 31.1 | 16 | 4.6 | 47 | 7:1 |
| *Leached* | 29.2 | 19 | 4.4 | 47 | 7:1 |
| *Podzolized* | 25.1 | 16 | 4.7 | 42 | 5:1 |
| *30–40 cm* | | | | | |
| *Calcareous* | 26.8 | 19 | 3.1 | 35 | 8:1 |
| *Common* | 31.0 | 17 | 3.3 | 42 | 9:1 |
| *Typical* | 31.4 | 15 | 4.2 | 48 | 7:1 |
| *Leached* | 29.7 | 17 | 4.3 | 43 | 9:1 |
| *Podzolized* | 27.8 | 42 | 4.3 | 43 | 6:1 |
| *60–70 cm* | | | | | |
| *Calcareous* | 22.9 | 28 | 3.4 | 34 | 7:1 |
| *Common* | 29.0 | 32 | 3.3 | 50 | 8:1 |
| *Typical* | 29.3 | 16 | 3.4 | 47 | 8:1 |
| *Leached* | 28.6 | 14 | 3.5 | 39 | 7:1 |
| *Podzolized* | 25.2 | 13 | 4.1 | 27 | 6:1 |
| *90–100 cm* | | | | | |
| *Calcareous* | 20.9 | 29 | 3.9 | 52 | 5:1 |
| *Common* | 19.5 | 32 | 3.4 | 51 | 5:1 |
| *Typical* | 26.8 | 17 | 2.6 | 50 | 8:1 |
| *Leached* | 26.6 | 17 | 2.6 | 25 | 10:1 |
| *Podzolized* | 25.5 | 31 | 4.1 | 41 | 6:1 |

## Reaction

A change in base saturation of the adsorption complex is reflected in the soil's *reaction*, which indicates whether it is alkaline, neutral of acidic. Reaction in water or in dilute salt solution is measured on the logarithmic pH scale – where neutrality is pH 7, the alkaline side of neutrality is shown by pH values above 7, and the acid side by pH values below 7.0. A completely desaturated soil will have a pH in the region of 4.0; a completely base-saturated soil will have a pH of 7.5–8.0 if free carbonate is present. Drawing on the standard reference (Krupenikov 1981), Table 2 summarizes the pH values for the topsoil of chernozem.

Most arable crops prefer a neutral to slightly acid reaction (pH 7.5–6.0) and the reaction of chernozem topsoils lies within these limits; the small coefficient

**Table 2** pH in chernozem at 0–40 cm depth

| Texture | Chernozem sub-type | Mean | Dispersion | Std dev | n |
|---|---|---|---|---|---|
| *pH in water* | | | | | |
| Heavy loam | *Calcareous* | 7.4 | 0.1 | 0.4 | 190 |
| Loam | *Calcareous* | 7.2 | 0.1 | 0.2 | 13 |
| Heavy loam | *Common* | 7.0 | 0.1 | 0.3 | 205 |
| Loam | *Common* | 6.9 | 0.1 | 0.4 | 52 |
| Heavy loam | *Xerophyte-wooded* | 7.0 | 0.1 | 0.4 | 47 |
| Heavy loam | *Common* | 6.7 | 0.1 | 0.3 | 171 |
| Loam | *Common* | 6.5 | 0.1 | 0.4 | 46 |
| Heavy loam | *Leached* | 6.6 | 0.1 | 0.3 | 174 |
| Heavy loam | *Typical* | 6.3 | 0.1 | 0.2 | 17 |
| Loam | *Leached* | 6.2 | 0.1 | 0.4 | 109 |
| Heavy loam | *Podzolized* | 6.9 | 0.1 | 0.4 | 77 |
| *pH in salt solution* | | | | | |
| Heavy loam | *Typical* | 6.5 | 0.1 | 0.3 | 35 |
| Heavy loam | *Leached* | 6.2 | 0.1 | 0.4 | 113 |
| Heavy loam | *Podzolized* | 6.1 | 0.1 | 0.3 | 71 |

of variation (3–6%) indicates that the pH range is practically constant. For the *Typical chernozem* (35 samples) pH in salt solution does not differ from pH in water but in *Podzolized chernozem* the difference between $pH_{water}$ and $pH_{salt}$ indicates the appearance of exchangeable acidity which replaces some of the bases in the adsorption complex.

Our overview of the stable features of the chernozem underscores its resilience in the face of processes of soil degradation. In respect of soil texture, mineralogy, the adsorption complex and reaction, the Moldavian chernozem is stable and very favourable to agriculture – and surpasses the chernozem of other regions (Alekseev 1999, Krupenikov 1987, Scerbacov and Vasinev 2000). Now we will consider some other soil attributes that are not so stable – and which need careful attention and active agricultural management.

# References

Alekseev VE 1999 *Mineralogy of soil formation in steppe and forest-steppe zones of Moldova: diagnosis, parameters, factors, processes*. Stiinta, Chisinau, 240p (Russian)

Fokin AD 1975 *Investigations of the processes of transformation, interaction and transmission of organic matter, iron and phosphorus in podzolic soils*. Dr Hab. Biological Sciences thesis, KA Timiriazev Agricultural Academy, Moscow, 350p (Russian)

Krupenikov IA (editor) 1981 *Statistical parameters for soil composition and properties in Moldova Vol.2*. Stiinta, Chisinau, 253p (Russian)

Krupenikov IA 1987 Chernozem of Europe and Siberia: similarities and differences. *Pochvovedenie* 11, 19–23 (Russian)

Scerbacov AP and Vasinev I 2000 *Anthropogenic evolution of chernozem soils*. Voronej State University, Voronej, 411p (Russian)

# Soluble Salts and Soil Solution

**Abstract** Excess of soluble salts in the soil impedes the uptake of water by roots. Chernozem are well drained so, in general, have low salinity. However, irrigation practice should avoid application of excess water so that a saline water table is not raised into the root zone. In chernozem, bicarbonate is the dominant anion in the soil solution but sulphates and chlorides may be present in the deep subsoil. Calcium is the dominant cation. The greatest natural loss of chemical elements in solution occurs in *Leached chernozem*, the least in *Calcareous chernozem*. Application of fertilizer increases the content of sulphur, calcium, magnesium and, especially, nitrate so there are consequent losses by deep percolation below the root zone.

**Keywords** Salinity · Soluble salts · Irrigation · Leaching

## Soluble Salts

Excess of soluble salts in the soil diminishes productivity by impeding rooting and water uptake, especially where the salt is close to the surface. Salinity generally occurs when drainage is restricted. Chernozem are generally well drained and so are generally free of excess soluble salts, at least in the topsoil. However, salinity is common in chernozem in parts of Siberia, the Tambov Plain, Northern Crimea, and the Hungarian Plain; other chernozem from these regions and, also, from the southern Ukraine and northern Caucasus, are salt free in the upper metre but contain excess soluble salts at greater depth.

We have collected analytical data for the salt content of four chernozem sub-types from Moldova to a depth of 2 m (Table 1). The analyses were performed on natural water extracted from dense natural aggregates under vacuum – a difficult procedure, so the 395 analyses represent no small effort. Nearly all the data are normally distributed and there is no significant difference in salt concentration between the sub-types or soil horizons.

The biggest salt concentration is 0.1 mg per 100 g of soil. Calcium is the dominant cation in the upper 2 m. Amongst the anions, bicarbonate ($HCO_3^-$) concentrations are twice those of sulphate ($SO_4^{2-}$). In general,

**Table 1** Soluble salt content of chernozem, mg/100 g of soil

| Depth (cm) | Mean mg/100 g | Std deviation | CoV (%) | n |
|---|---|---|---|---|
| *Calcareous* | | | | |
| 0–50 | 0.07 | 0.02 | 33 | 29 |
| 50–100 | 0.07 | 0.02 | 38 | 30 |
| 100–150 | 0.06 | 0.02 | 36 | 22 |
| 150–200 | 0.06 | 0.02 | 34 | 23 |
| *Common* | | | | |
| 0–50 | 0.08 | 0.03 | 36 | 38 |
| 50–100 | 0.08 | 0.03 | 33 | 35 |
| 100–150 | 0.08 | 0.03 | 34 | 17 |
| 150–200 | 0.08 | 0.02 | 25 | 12 |
| *Typical* | | | | |
| 0–50 | 0.07 | 0.03 | 38 | 33 |
| 50–100 | 0.08 | 0.03 | 30 | 39 |
| 100–150 | 0.09 | 0.03 | 30 | 15 |
| 150–200 | 0.09 | 0.02 | 26 | 11 |
| *Leached* | | | | |
| 0–50 | 0.08 | 0.02 | 26 | 35 |
| 50–100 | 0.08 | 0.03 | 38 | 31 |
| 100–150 | 0.11 | 0.05 | 46 | 12 |
| 150–200 | 0.08 | 0.04 | 44 | 13 |

absolute concentrations are not high, although ample from the point of view of plant nutrition (Ursu 1984, p 135).

As a rule of thumb, sensitive crops are affected by salinity greater than 0.2 mg/100 g soil. Generally, Moldavian chernozem contain less than half of this amount but soluble salts are sometimes present in the deep subsoil. Sodium salts, in particular, are toxic and bad for soil structure and permeability. For this reason, only low-salt water should be used for irrigation. Application of water should not exceed what is needed over and above natural rainfall to supply crop water requirements and to leach surplus salts from the root zone (Zimovet 1991); the depth of wetting must not be more than 2 m or soluble salts may be brought up from greater depth by capillary action.

## Soil Solution

The standard determination of salts in a water extract, using a soil:water ratio of 1:5, is convenient but gives a distorted picture of the salt concentration in the soil. The real picture is given by soil solution extracted directly from lysimeters. This has been studied over many years by Z. Sinkevici (1989) who found that in chernozem:

1. Soil solutions have a neutral or slightly alkaline reaction. In *Leached chernozem* and *Typical chernozem*, the pH of solutions from the topsoil is about 6.5; in *Calcareous chernozem* pH is from 7.1 to 7.5 in the topsoil and 7.1 to 8.5 in the deeper layers
2. Total salt concentration is only 0.1–0.3 g/L
3. Calcium and magnesium are the main cations; silica concentration is low; aluminium, iron and phosphorus very low or not detectable; the main anions are $HCO_3^-$ and $SO_4^{2-}$ with very little chloride and nitrate
4. The degree of soil wetting varies with the season and from year to year
5. The greatest leaching occurs in the *Leached chernozem* and least in the *Calcareous chernozem*

Sinkevici also studied the influence of mineral fertilizers on the composition of soil solution. Application of fertilizer brings a 2-fold to 4-fold increase in sulphate, calcium and magnesium, and a 12-fold increase in nitrate in the topsoil. There are also increased losses by leaching below the root zone. We shall return to this issue but should note here that nitrogenous fertilizers should be introduced carefully to avoid nitrates leaching through the soil and polluting the groundwater.

## References

Sinkevici ZA 1989 *Modern processes in chernozemic soils of Moldova*. Stiinta, Chisinau, 215p (Russian)

Ursu AT (editor) 1984 *Soils of Moldova. Vol.1 Genesis, ecology, classification and systematic description of soils*. Stiinta, Chisinau, 351p (Russian)

Zimovet BA 1991 *Ecology and improvement of soils in the drought zone*. Nauk Moscow (Russian)

# Humus – Guardian of Fertility and Global Carbon Sink

**Abstract** Humus is the defining constituent of soil, the mainspring of soil fertility and the outstanding feature of the chernozem. The quality of humus in chernozem is uniquely favourable to agriculture, with a preponderance of humic acids bound up with calcium. Eastwards, across the Ukraine and Russia, chernozem keep their humus but not the same great fertility – because of the increasingly severe continental climate. Quantitatively, the main elements in living organisms are carbon, nitrogen, oxygen, hydrogen, phosphorus and sulphur. The same is true of humus – so decomposition of humus provides plants with all essential nutrients. But, in farming systems, nutrients taken up by the crops are carried away so, without the renewal of humus reserves, farming is not sustainable. Soils are the largest pool of terrestrial organic carbon; estimates range between 1000 and 3500 billion tonnes. Between one-quarter and one-third of the excess carbon dioxide in the atmosphere, which is considered to be driving climatic change, has come from land use change over the last century. Annual carbon emissions from chernozem are of the order of 200 million tonnes, a little over 2% of today's global emissions, but farmland represents the outstanding opportunity to fix carbon and put it back where it came from – in the soil, where it will be useful! Comparison of today's humus stocks with those measured at the end of the nineteenth century shows a dramatic decline following the ploughing of the steppe. It might have been expected that the humus content would stabilize at a new equilibrium but this has not happened; over the last 40 years, the humus content has declined by about 0.3%/year. Over the last 100 years, the loss of humus from the 0 to 30 cm layer has been in Moldova 51–71 t/ha (32–40% of the original stock or 0.5–0.7 t/ha/year); in the Kursk and Kharkov regions of Russia and Ukraine 67–79 t/ha (21–36% of the original stock or 0.7–0.8 t/ha/year); and in the Samara region 150–180 t/ha (38–39% of the original stock or 1.5–1.8 t/ha/year). Recent data show equally dramatic decline of humus content below the plough layer to a depth of 120–130 cm, especially under bare fallow. Calculation of data in terms of equivalent soil mass gives less dramatic but probably more realistic values for these losses: for *Typical chernozem* near Kursk, a loss of 40.4 t/ha of carbon and 3.5 t/ha of nitrogen after a century of arable cropping; 76.9 t/ha of carbon and 4.5 t/ha of nitrogen after 50 years of continuous bare fallow. At the high noon of chemical agriculture the balances for nitrogen, phosphate and potassium were positive. This decreased but did not eliminate the loss of humus. Nutrient balances are again declining sharply which implies a long-term decline in productivity; food security is not assured – and by 2025 Moldova will no longer be a country of chernozem.

**Keywords** Humus · Soil organic carbon and nitrogen · Soil fertility · Carbon stocks · Fixation and emissions · Humus quality

## Its Significance for Farming

Humus is surely the defining constituent of soil, the symbol and mainspring of soil fertility – and the outstanding feature of the chernozem. Echoing P.A. Kosticev, Viliams (1940) pronounced that the essence of soil formation is "the synthesis and the decomposition of soil organic matter". Humus influences so many features of the soil: its colour, structure, many physicochemical and biochemical functions and, especially, the supply of plant nutrients. There is a huge literature on humus but all its secrets are not yet discovered.

For the Roman agronomists Virgil and Columnella, *humus* meant soil. The word was gradually abandoned and replaced by *terra* but the Latin word reappears in the eighteenth century in its modern meaning when the *humus theory* of plant nutrition was proposed by J.A. Külbel in 1741 and elaborated by A. von Thaer. Thaer was the first to define humus precisely as a soil constituent: "The usual name for this substance is mould... Humus is the residue of animal and plant putrefaction" (Thaer 1811–1816, vol. 2, pp 102–114). According to Thaer "As humus is a life creation, it also creates life; it feeds organic bodies. Life is impossible without it ... Thus death and decomposition are necessary for new life." His reasoning was that the natural turnover of organic matter is a closed system, quantitatively unchanged and with no relations between the mineral world and the organic world – so the essence of life is the *life–death* or, strictly, *life–lifeless* cycle. The theory was backed up by a mass of data, including research of his own and his associates, and supported by such high authorities as Humphry Davy in England, J.J. Berzelius in Sweden and M. Pavlov – professor of Moscow University who won the nickname of Russian Thaer (Viliams 1940). Thaer's mistake was to base his remarkable work of data collection and reflection on soil management on a wrong hypothesis – that humus is the principle source of carbon for the plant (Feller et al., 2003). However, Liebig's proposition of the mineral nutrition of plants in 1840 (Liebig 1840, 1851) drew much from Thaer – and there is more than a little of the same metaphysics in today's fashion for organic farming.

Quantitatively, the main elements of life are carbon, nitrogen, oxygen, hydrogen, phosphorus and sulphur; and so it is with humus – the decomposition of humus provides plants with all essential nutrients. However, in farming systems nutrients are carried away with the crop and, without the renewal of this reserve, a deficit is inevitable.

Since carbon is the largest component of humus, we can measure the decomposition and replenishment of humus most easily in terms of carbon cycling. Soils are the largest pool of terrestrial organic carbon but the depiction of the global carbon cycle in Fig. 1 probably underestimates the soil's contribution. Few estimates have been made using a large number of representative samples or for depths greater than 1 m. The more plausible indicate a range of 1462–1548 billion tonnes of soil organic carbon in the topmost metre, 2376–2456 in the top 2 m (Batjes 1996) and 2344 billion tonnes in the upper 3 m (Jobbagy and Jackson 2000). These are all rough estimates because there is poor knowledge of the global pattern of soils and land use, and a lack of detailed data on the vertical distribution of soil organic matter, bulk density and stoniness. Importantly, especially in the case of the chernozem, most estimates have been limited arbitrarily to a depth of 1 m. Glazovskaya (1996) evaluated the vertical distribution of organic carbon in *Calcareous, Ordinary, Typical* and *Podzolized chernozem* in Russia, using 112 profiles, and found that the upper 0.5 m depth, for which most data are available, contains only 45–66% of the organic carbon. She proposed that the current estimates of carbon storage in chernozem should be increased by 20%. More recently Mikhailova and Post (2006), using a more sophisticated model, estimate that 27–30% of the soil organic carbon in the *Typical chernozem* occurs below 1 m.

We may note at this point that the soil's nitrogen content is strongly related to humus; nitrogen makes up about 5% of soil organic matter (see Chapter 8).

All the mineral components of organic matter return to the soil through litter-fall but, in farming systems, a deficit is inevitable because part of them is removed with the crop and needs to be replaced by fertilizer or manure. We have already mentioned that the history of agriculture is full of paradoxes; and another is that, in comparison with the much-derided "period of stagnation", the use of mineral fertilizers in Moldova has *decreased* 20-fold. Many fields have not seen fertilizer for a long time and crops now depend entirely on humus – that is, on the decomposition of the humus. Of course, there is also humus decomposition in virgin chernozem but, simultaneously, renewal prevails so, over a long period, every hectare accumulated hundreds or even thousands of tons of humus. Nowadays, while this gold reserve is yielding crops, it is wasting. This is a tragedy – and not only for the soil and ecosystems. It is an economic loss and a loss to our society. We should not be deceived by the relatively good yields at the turn of the millennium – favourable weather and remnants of previous fertilizer applications contributed, and

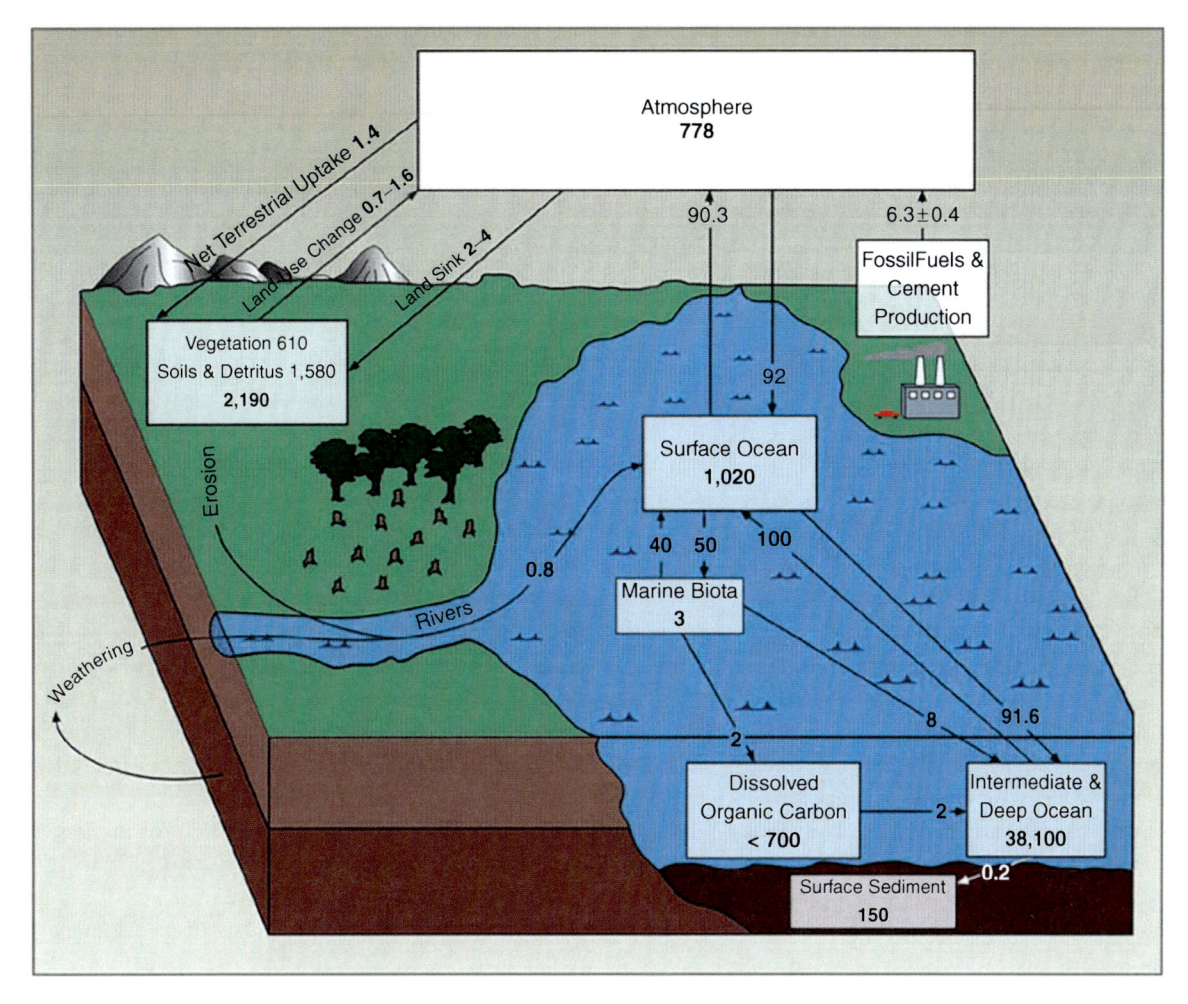

**Fig. 1** The global carbon cycle (IGPCC 2010) Carbon pools in petagrams carbon (PgC = 10 to power 9 tonnes) and fluxes in PgC/yr

even those yields were inferior to those of the 1970s and 1980s.

## The Global Carbon Sink: Soils and Climatic Change

The loss of humus is not just an issue of soil fertility. Carbon dioxide is the principal greenhouse gas in the atmosphere and its increasing concentration is considered to be driving climatic change. Between one-quarter and one-third of the excess carbon dioxide in the atmosphere has come from land use change over the last century; it has been emitted by land cleared from forest or grassland and from the continuing decomposition of humus in arable soils. Over this period, the chernozem with its enormous humus reserve has been a significant source – rather than a sink – for carbon. If we consider the global extent of chernozem (230 million hectares), most of which is now cultivated, and a humus stock of 360–450 t of carbon per hectare under native grassland, and taking an average loss of 33% of the initial humus stock from the top metre, we arrive at total emissions of 19–24 billion tonnes of carbon.[1]

---

[1] Assuming that 58% of reported humus values are organic carbon (Brady and Weil 1996).

**Table 1** Humus balance (t/ha) and nutrient balance (kg/ha) in Moldavian soils

| Period | Humus | N | $P_2O_5$ | $K_2O$ |
|---|---|---|---|---|
| 1961–1965 | −1.5 | −2.7 | −6 | −28 |
| 1966–1970 | −1.2 | −2.6 | −2 | −35 |
| 1971–1975 | −0.8 | −25 | +4 | −35 |
| 1978–1980 | −0.7 | −7 | +20 | −14 |
| 1981–1985 | −0.4 | +24 | +35 | +24 |
| 1986–1990 | −0.3 | +16 | +35 | +17 |
| 1991–1995 | −0.8 | −18 | −11 | −80 |
| 1996–2000 | −1.1 | −36 | −51 | −150 |

Global carbon emissions have risen from 4 billion tonnes in 1960 to close on 10 billion tonnes a year now, as a result of the increased use of fossil fuels (le Quéré et al. 2009). The estimated land use change component remains at about 1.5 billion t/year. Taking a conservative figure for the annual loss of humus of 0.7 t/ha for the upper 30 cm of chernozem under arable crops (Table 1) to represent an annual loss of 1.4 t/ha over the whole soil profile (equivalent to 0.82 t of carbon), chernozem globally account for annual carbon emissions of 203 million tonnes. This was almost 5% of global carbon emissions in the 1960s, when land use change was a relatively greater contribution of carbon dioxide compared with today, and a little over 2% of today's much greater global emissions. At the same time, the chernozem represents an outstanding opportunity to fix carbon and put it back where it came from – in the soil, where it will be useful!

How to do this is the subject of the final part of this book. In short, ecological agriculture is a sustainable way to rebuild the rural economy and combat climatic change.

## Humus Stocks and Trends

It is hard to assess the stock of humus in Moldavian chernozem because investigations have been curtailed. We have to piece together the picture from disparate sources that include the book *Moldavian Chernozem* (Krupenikov 1967), more detailed data in the monograph *Moldavian Soils* (Ursu 1984), and a map of humus composition in the *Atlas of the Moldavian SSR* (GVGK 1978) which estimates that the topmost metre contains 1 billion tonnes of humus and 50 million tonnes of nitrogen. Many useful data were compiled in *Statistical Parameters of Moldavian Soils* (Krupenikov 1981) and repeated or summarized in the *Bank of Data* published in 2000 (Cerbari 2000). Table 2 takes an example of the humus and nitrogen content in four chernozem sub-types and Table 3 presents data on the mean humus composition of Moldavian soils based on 1978 statistics.

The deep and regular penetration of humus is most characteristic of *Typical chernozem*; the C:N ratio in the upper 20 cm is about 10 and the ratio narrows in the deeper layers, so deeper humus is richer in nitrogen than topsoil. All Moldavian chernozem contain more than 1% humus at the depth of 110–120 cm, and this is the source of their fertility. Further east, across the Ukraine and Russia, the black

**Table 2** Humus and total nitrogen in heavy loam chernozem, % (total samples: humus = 690, nitrogen = 1527)

| Depth (cm) | Calcareous | | | Common | | | Typical | | | Leached | | |
|---|---|---|---|---|---|---|---|---|---|---|---|---|
| | Humus | N | C:N | Humus | N | C:N | Humus | N | C:N | Humus | N | C:N |
| 0–20 | 4.05 | 0.22 | 10.2 | 3.98 | 0.21 | 10.6 | 4.35 | 0.24 | 11.3 | 4.24 | 0.23 | 11.1 |
| 20–30 | 3.33 | 0.18 | 10.2 | 3.68 | 0.20 | 10.5 | 3.93 | 0.23 | 10.7 | 3.89 | 0.22 | 10.9 |
| 30–40 | 3.30 | 0.19 | 10.3 | 3.40 | 0.19 | 10.5 | 3.58 | 0.21 | 10.8 | 3.50 | 0.20 | 10.8 |
| 40–50 | 2.74 | 0.17 | 9.8 | 2.82 | 0.17 | 9.9 | 3.09 | 0.18 | 10.9 | 2.96 | 0.17 | 10.9 |
| 50–60 | 2.69 | 0.17 | 9.2 | 2.44 | 0.15 | 9.7 | 2.57 | 0.12 | 11.1 | 2.38 | 0.16 | 10.7 |
| 60–70 | 2.11 | 0.14 | 9.2 | 1.95 | 0.14 | 8.7 | 2.20 | 0.15 | 11.2 | 2.16 | 0.15 | 10.8 |
| 70–80 | 1.94 | 0.09 | 8.6 | 1.74 | – | – | 1.71 | – | – | 1.69 | 0.11 | 9.2 |
| 80–90 | 1.46 | 0.10 | 8.1 | 1.38 | 0.10 | 8.2 | 1.47 | – | – | 1.45 | 0.9 | 8.8 |
| 90–100 | 1.26 | – | – | 1.14 | – | – | 1.16 | 0.08 | 8.7 | 1.08 | – | – |
| 100–110 | 1.06 | – | – | 1.04 | – | – | 1.01 | – | – | 1.06 | – | – |
| 110–120 | 0.98 | – | – | 0.95 | – | – | 0.79 | – | – | 0.83 | – | – |
| 120–130 | 0.82 | – | – | 0.92 | – | – | 0.78 | – | – | 0.80 | – | – |
| 130–140 | 0.64 | – | – | 0.79 | – | – | 0.86 | – | – | 0.78 | – | – |
| 140–150 | 0.46 | – | – | 0.73 | – | – | 0.86 | – | – | 0.83 | – | – |

**Table 3** Mean humus content in Moldavian soils

| Soil type and sub-type | Humus (%) | |
|---|---|---|
| | 0–50 cm | 50–100 cm |
| *Calcareous chernozem* | 3.36 | 1.65 |
| *Common chernozem* | 3.47 | 1.73 |
| *Xerophyte-wooded chernozem* | 6.04 | 2.15 |
| *Xerophyte-wooded chernozem* (ploughland) | 3.99 | 1.99 |
| *Typical chernozem* | 3.74 | 1.83 |
| *Leached chernozem* | 3.40 | 1.95 |
| *Podzolized chernozem* (ploughland) | 3.42 | 1.62 |
| *Dark grey forest* (ploughland) | 2.49 | 1.09 |
| *Grey forest* (ploughland) | 1.82 | 0.72 |

earths keep their humus but not the same great fertility – because of the increasingly severe continental climate (Krupenikov 1987).

The old figures cannot be compared directly with modern data: there are no modern data but Ion Burlacu's maps using data compiled for the 5-year plans from 1966–1970 until 1991–1995 (Burlacu 2000) provide information for the top 30 cm layer; extrapolation may indicate the rates of change over the last 2 decades. The content of humus in the 0–30 cm layer declined from 4–5% in 1950 to 3.5–4% in 1965 and 3.0–3.5% in 1990 (Ursu 2000). During this period, the amount of farmyard manure applied was of the order of 6–8 t/ha/year.

The substantial application of mineral fertilizer brought benefits but, also, damage. In Moldova, 400,000 ha of farmland were stripped of phosphate, corn crops suffered chlorosis and needed zinc supplements, and excess nitrate was leached to the groundwater and the rivers.

By extrapolating the trend of humus decrease, Andries and Zagorcea (2002) predicted a long-term decline in productivity (Table 4). We should be sceptical about any prognosis but this one is quite probable and the situation in 2025 might be worse; food security is not assured and Moldova will no longer be a country of chernozem.

## Losses of Humus Under the Plough

Even before the systematic studies of humus dynamics by Krupenikov (1967), Korduneanu (1985) and Zagorcea (1990), it was observed that Moldavian soils had lost about 60 t of humus per hectare during the first 80–100 years under the plough. In Chapter 1, we mentioned that the founder of pedology, V.V. Dokuchaev, carried out fieldwork in Bessarabia in 1877. In his book *Russian Chernozem* (1883) he wrote "from Iampol to Soroca, I followed the right bank of the Dniester through the property of N.K. Erju – Nepada." He describes the alluvium by the waterside and, away from the steep bank of the river, deep colluvial layers, then "In order to see the normal soils of Bessarabia ... I made a trip to Cuhuresti which is located 12–14 versts to the south west. On the upper terrace of the Dniester, 4–6 verst (5–7 km) from Nepada, in the virgin steppe, absolutely plain, the soil profile revealed:

A – Soil horizon penetrated by roots, living and dead, the soil absolutely black and friable – 61 cm;
B – Transitional horizon with burrows and casts of the blind mole-rat –30 cm;
C – Typical loess with many concretions."

"The chernozem seemed to me in 1877 so typical that I called it first class. The analysis showed the content of humus was 5.718%. The same soil was found in the country around Cuhuresti." The chernozem of Soroca is represented as the westernmost extent of Dokuchaev's *Typical chernozem:* "We would like to

**Table 4** Relationships between soil humus and nitrogen and crop yields with forecasts to 2025

| Year | Humus (%) | Humus stock 0–30 cm (t/ha) | | Yield of field crops (t/ha) | Mineral nitrogen applied (kg/ha) | |
|---|---|---|---|---|---|---|
| | | Humus | Nitrogen | | Winter wheat | Corn |
| 1897 | 5–6 | 200 | 10 | 1.35 | – | – |
| 1950 | 4–5 | 150 | 8 | 1.15 | – | – |
| 1965 | 3.5–4 | 130 | 6 | 1.05 | 32 | 42 |
| 1990 | 3.0–3.5 | 110 | 5 | 0.85 | 25 | 34 |
| 2025 | 2.5–3.0 | 90 | 4 | 0.70 | 21 | 28 |

**Table 5** Comparison of Dokuchaev's *Typical chernozem* with modern observations

| Index | | 1877 steppe | 1960 arable | | 2003 | |
|---|---|---|---|---|---|---|
| | | | Dniester terrace | Plateau | arable | Shelter belt |
| Soil profile morphology, depths in cm | A | 0–61 | 0–43 | 0–44 | 0–50 | 0–50 |
| | B | 62–91 | 44–101 | 45–92 | 51–98 | |
| | C | 92+ | 102+ | 93+ | 99+ | |
| | Effervescence | No data | 92 | 65 | 70 | No data |
| Humus content, % | 0–20 | 5.72 | 3.75 | 3.60 | 3.36 | 4.2 |
| | 30–40 | | 3.65 | 3.30 | 3.15 | |
| | 50–70 | | 2.34 | 2.73 | 1.94 | |
| | 70–90 | | 1.59 | 1.57 | 1.68 | |

mention the content of humus, 5–6% as typical for the virgin chernozem" (not ploughed).

During systematic soil survey of the country around Soroca in 1955–1960, A. Ursu retraced the footsteps of Dokuchaev and sampled two soil profiles towards Cuhuresti – one in the same place as the original observations on the flat upper terrace of the Dniester, 4–6 versts (5–7 km) from Nepada and another on the plateau towards Cuhuresti. In Dokuchaev's day, this was all pristine steppe; in 1960 it was ploughland and had been cultivated for at least 60 years. Ursu repeated the analysis in 2003 with three replicates at the arable site on the Dniester terrace and, also, in a nearby woodland shelter belt (Ursu 2005). He describes the arable soil, comparable to Dokuchaev's original observation, as

Ap 0–25 cm Dark grey (dry) clay, weak granular structure, compacted plough layer;
Ah 25–50 cm Dark grey (moist) clay, fine-granular structure;
B1 50–70 cm Dark brownish grey, clay, compacted, granular structure weakly developed, carbonates present below 70 cm;
B2 75–98 cm Parti-coloured brown and yellow, well structured;
C 98–150 cm Yellow loess-like clay with white carbonate streaks.

Table 5 compares the historical observations – although it is hard to make direct comparison because bulk density was not measured on either occasion; the conventional determination of humus as per cent by mass cannot be converted to the three-dimensional soil layers in the landscape. There appears to be a reduction in thickness of the A horizon, from 61 to 43/50 cm, although the total thickness of the solum (A+B) does not vary much. What we do see is a dramatic decline in humus content after the ploughing of the steppe. It might have been expected that the humus content would stabilize at a new equilibrium but this did not happen; over the last 40 years, the humus concentration declined in absolute terms by about 0.3% annually in the plough layer and by the same magnitude throughout the A and B horizons. By contrast, the humus content increased beneath the grassed and wooded shelter belt, established since 1960. We return to the issue of rebuilding the humus status in Part III.

This is no isolated case: *Russian Chernozem 100 Years After Dokuchaev* (Kovda 1983) makes a striking comparison between the Dokuchaev's iso-humus lines and modern data (Table 6).

For the 0–30 cm layer, the loss of humus in Moldova was 51–71 t/ha (32–40% of the original stock or 0.5–0.7 t/ha/year); in the Kursk and Kharkov regions of Russia and the Ukraine, the loss was 67–79 t/ha (21–36% of the original stock or 0.7–0.8 t/ha/year); and in the Samara region 150–180 t/ha (38–39% of the original stock or 1.5–1.8 t/ha/year). A total loss of 30–40% of the initial stock is typical. A similar picture is revealed by a systematic review of published studies of change in soil organic matter in Canada, which showed a loss of 17–58% (mean 30) of soil organic carbon from black earth soils after native grassland was converted to arable (VandenBygaart et al., 2003); earlier studies on chernozem in the Prairie Provinces recorded losses of 22±4 and an average of 24% loss of soil organic carbon (Newton et al., 1945, Reinl 1984).

There have also been great losses from deeper layers but comparable data are scarce. Mikhailova et al. (2000), in the Kursk area of Russia, have compared *Typical chernozem* on pristine steppe, uncultivated for

**Table 6** Change in humus content of the plough layer (0–30 cm) of chernozem from the European USSR over 100 years

| Chernozem sub-type | Region | Humus content and stock | | | | Humus loss over 100 years (t/ha) | Mean loss (t/year) | Humus loss, % initial stocks |
|---|---|---|---|---|---|---|---|---|
| | | 1881 | | 1981 | | | | |
| | | % by mass | t/ha | % by mass | t/ha | | | |
| *Typical* | Tambov, Voronezh | 10–13 | 300–390 | 7–10 | 210–300 | 90 | 0.9 | 23–30 |
| | Kursk, Kharkov | 7–10 | 221–315 | 4–7 | 142–248 | 67–79 | 0.7–0.8 | 21–36 |
| *Leached* | Stavropol | 7–10 | 231–330 | 4–7 | 150–263 | 67–81 | 0.7–0.8 | 20–34 |
| *Common* | Voronezh | 7–10 | 221–315 | 4–7 | 150–263 | 52–71 | 0.5–0.7 | 17–32 |
| | Moldova | 4–7 | 126–221 | 2–4 | 75–150 | 51–71 | 0.5–0.7 | 32–40 |
| *Typical* | Samara (Kuybyshev) | 13–16 | 390–480 | 8–10 | 240–300 | 150–180 | 1.5–1.8 | 38–39 |
| *Common* | Orenburg | 9–11 | 270–330 | 6–8 | 180–240 | 90 | 0.9 | 27–33 |
| *Leached* | Ulianovsk | 13–16 | 390–480 | 4–7 | 120–210 | 270 | 2.7 | 56–69 |

at least 300 years, with comparable sites under arable cultivation for at least 100 years and with a 50-year continuous bare fallow. Using a statistically rigorous sampling design, they collected samples every 10 cm to a depth of more than 2 m and, importantly, also measured bulk density. Compared with pristine steppe, the cultivated soils show losses of humus to a depth of 120 cm. On the conventional basis of measured concentrations by mass of the sample, the continuous arable shows losses of 17–38% of organic carbon and 15–45% of the total nitrogen; the bare fallow shows even greater losses of 26–43% organic carbon and 25–53% of total nitrogen.[2] Table 7 is re-calculated in terms of equivalent soil mass (Ellert and Bettany 1995), which allows for the simultaneous changes in soil bulk density resulting from a collapse of soil structure as humus is lost. The changes appear less dramatic than comparison of raw sample concentrations but may be more valid and ecologically relevant. Compared with pristine steppe, the soil after a century of continuous cropping shows total losses of 40.4 t/ha of carbon and 5.48 t/ha of nitrogen, with depletion to a depth of 60 cm. After 50 years of continuous bare fallow, the soil is depleted in organic carbon and total nitrogen to an equivalent depth of 1 m, with total losses of 76.94 t/ha of carbon and 8.48 t/ha of nitrogen. This demonstrates that the inputs of organic matter from arable crops do not match the losses; and under bare fallow there is no recompense at all.

## The Quality of Humus

New data on soil humus, obtained by modern methods, will be analysed in Part III. Now, we consider the forms of humus that turn over relatively quickly in the soil (the so-called labile fraction) and their role in plant nutrition. Here we are using the old-fashioned interpretation in terms of *humic* and *fulvic* acids that are determined by separating organic matter from the mineral fraction by treatment with sodium hydroxide solution.[3] The alkaline extract is acidified, which precipitates some of the dispersed organic matter. That part which stays in solution is known as *fulvic acid*; that which precipitates is known as *humic acid*. A further fraction of organic matter that does not disperse in sodium hydroxide is known as *humin*.

By conventional interpretation, humic acids are of high quality – being associated with calcium and having a higher nitrogen content than fulvic acids (Ganenco 1991). The humic acid:fulvic acid ratio ($C_h:C_f$) is taken as an index of the quality of the organic matter. Comparing the index for Moldavian soils (Table 8), the various sub-types of chernozem (profiles 1–10) are close together: the carbon content decreases with depth and the $C_h:C_f$ ratio ranges

---

[2]Most measurements of soil organic carbon reported here were obtained by Tyurin's method (Bel'chikova 1954) which tends to underestimate carbon values compared with modern determinations by dry combustion (Sorokina and Kogut 1997) used by Mikhailova et al. 2000. A conversion factor of 1.13 may be applied to arrive at equivalent values.

[3]As originally determined by Bertholet and André (1892).

**Table 7** Comparison of soil organic carbon and nitrogen contents in *Typical chernozem* between native grassland, arable and bare fallow. After Mikhailova and others 2000

| Thickness (cm) | Bulk density (t/m³) | Equiv. soil mass (t/ha) | Organic C mass (t/ha) | Total N mass (t/ha) | Organic C loss (t/ha) | Total N loss (t/ha) |
|---|---|---|---|---|---|---|
| Native grassland | | | | | | |
| 0–14.6 | 0.8±0.09 | 1170.0 | 64.75±5.43 | 6.35±.60 | | |
| 14.6–27.1 | 0.94±0.07 | 1171.8 | 52.12±4.30 | 4.82±.40 | | |
| 27.1–39.5 | 0.94±0.07 | 1171.8 | 46.36±3.96 | 4.02±.31 | | |
| 39.5–51.5 | 1.01±0.08 | 1210.0 | 44.21±5.62 | 3.85±.46 | | |
| 51.5–63.2 | 1.06±0.04 | 1240.0 | 31.33±4.56 | 2.63±.46 | | |
| 63.2–73.2 | 1.08±0.04 | 1060.0 | 30.18±1.99 | 2.47±.24 | | |
| 73.2–82.8 | 1.11±0.04 | 1080.0 | 28.15±2.09 | 2.26±.20 | | |
| 82.8–92.8 | 1.11±0.04 | 1110.0 | 21.5±2.07 | 1.80±.15 | | |
| 92.8–102.6 | 1.11±0.04 | 1090.0 | 20.54±3.68 | 1.72±.31 | | |
| 102.6–112.2 | 1.15±0.04 | 1110.0 | 19.78±3.99 | 1.63±.32 | | |
| 112.2–121.7 | 1.16±0.04 | 1110.0 | 13.80±2.65 | 1.21±.21 | | |
| 121.7–131.4 | 1.20±0/04 | 1150.0 | 12.70±3.45 | 1.15±.30 | | |
| Continuously cropped | | | | | | |
| 0–10 | 1.17±0.09 | 1170.0 | 40.34±3.57 | 3.51±0.30 | 24.41 | 2.84 |
| 10–19.5 | 1.24±0.10 | 1171.8 | 40.17±3.49 | 3.41±0.30 | 11.95 | 1.41 |
| 19.5–28.9 | 1.24±0.10 | 1171.8 | 38.51±3.35 | 3.21±.028 | 7.85 | 0.81 |
| 28.9–39.0 | 1.20±0.12 | 1210.0 | 40.34±5.36 | 2.80±0.44 | NS | NS |
| 39.0–49.3 | 1.21±0.11 | 1240.0 | 28.58±3.41 | 2.43±0.29 | NS | NS |
| 49.3–59.3 | 1.06±0.04 | 1060.0 | 23.68±3.23 | 2.05±0.33 | 6.5 | 0.42 |
| 59.3–68.0 | 1.08±0.04 | 1080.0 | 23.65±2.87 | 2.11±0.17 | NS | NS |
| 68.0–78.0 | 1.11±0.04 | 1110.0 | 19.19±2.73 | 1.79±0.24 | NS | NS |
| 78.0–88.0 | 1.09±0.04 | 1090.0 | 18.64±1.72 | 1.65±0.17 | NS | NS |
| 88.0–98.0 | 1.11±0.04 | 1110.0 | 16.52±2.59 | 1.52±0.22 | NS | NS |
| 98.0–108.0 | 1.15±0.04 | 1110.0 | 13.99±2.03 | 1.31±0.17 | NS | NS |
| 108.0–118.0 | 1.16±0.04 | 1150.0 | 13.04±3.12 | 1.24±0.26 | NS | NS |
| 50-year continuous bare fallow | | | | | | |
| 0–10.7 | 1.09±0.12 | 1170.0 | 37.04±3.27 | 3.00±0.28 | 27.71 | 3.35 |
| 10.7–20 | 1.26±0.11 | 1171.8 | 39.54±4.30 | 3.18±0.35 | 12.58 | 1.64 |
| 20–30 | 1.26±0.11 | 1171.8 | 39.34±4.41 | 3.15±0.33 | 7.02 | 0.87 |
| 30–40 | 1.21±0.11 | 1210.0 | 35.33±5.36 | 2.82±0.41 | 8.88 | 1.03 |
| 40–50 | 1.24±0.15 | 1240.0 | 30.65±6.32 | 2.44±0.42 | N S | N S |
| 50–60 | 1.06±0.04 | 1060.0 | 23.11±3.84 | 1.83±0.22 | 7.07 | 0.64 |
| 60–70 | 1.08±0.04 | 1080.0 | 20.97±3.48 | 1.74±0.20 | 7.18 | 0.52 |
| 70–80 | 1.11±0.04 | 1110.0 | 18.87±1.98 | 1.60±0.17 | NS | NS |
| 80–90 | 1.09±0.04 | 1090.0 | 15.59±2.84 | 1.35±0.16 | NS | NS |
| 90–100 | 1.11±0.04 | 1110.0 | 13.28±2.08 | 1.20±0.13 | 6.50 | 0.43 |
| 100–110 | 1.15±0.04 | 1110.0 | 12.85±1.49 | 1.15±0.11 | NS | NS |
| 110–120 | 1.16±0.04 | 1150.0 | 10.56±1.47 | 1.01±0.09 | NS | NS |

from 1.8 to 3.4 in the topmost 25 cm to 2.5–5.6 at 50–60 cm; the compact varieties have a special character with high $C_h:C_f$ ratios, except in the topsoils which are similar to normal chernozem. In forest soils, the $C_h:C_f$ ratio is narrower and fulvic acids may even be predominant (Ursu 1984, Grati 1977).

Figure 2 depicts the composition of humus in the same representative soil profiles. There is a very small proportion of inert *bitumen* and other *recalcitrant* organic matter, except in forest soils; *humin* comprises about 20%, rather more at depth; and the preponderance of *humic acids*, associated with calcium, in the chernozem again testifies to the quality of their humus.

Apart from humus, that is soil organic carbon, carbon occurs in soil as calcium and magnesium

**Table 8** Humus composition of Moldavian soils

| | Soil type | Depth (cm) | Total C % | Carbon, % total soil C | | Humic:fulvic |
| --- | --- | --- | --- | --- | --- | --- |
| | | | | Humic acids | Fulvic acids | |
| 1 | *Leached chernozem* | 0–20 | 2.64 | 50.1 | 19.3 | 2.6 |
| | | 30–40 | 2.30 | 43.3 | 19.6 | 2.2 |
| | | 60–70 | 1.57 | 44.9 | 18.8 | 2.3 |
| 2 | *Leached chernozem* | 0–20 | 3.30 | 48.4 | 16.3 | 3.0 |
| | | 30–40 | 2.93 | 49.1 | 17.8 | 2.8 |
| | | 50–60 | 1.40 | 41.7 | 18.8 | 2.2 |
| 3 | *Typical chernozem* (deep calcareous layer) | 0–27 | 2.36 | 45.3 | 21.5 | 2.1 |
| | | 27–47 | 2.06 | 39.8 | 20.9 | 1.9 |
| | | 47–57 | 1.18 | 38.9 | 18.5 | 2.1 |
| 4 | *Typical chernozem* | 0–10 | 2.92 | 45.3 | 24.7 | 1.8 |
| | | 25–35 | 2.45 | 42.0 | 20.0 | 2.1 |
| 5 | *Common chernozem* | 0–20 | 2.70 | 50.8 | 14.8 | 3.4 |
| | | 30–40 | 2.33 | 49.6 | 19.4 | 2.6 |
| | | 50–60 | 1.64 | 47.6 | 20.8 | 2.2 |
| | | 90–100 | 0.77 | 44.5 | 20.9 | 2.1 |
| | | 310–320 | 0.49 | 46.1 | 14.6 | 3.1 |
| 6 | *Common chernozem* | 0–20 | 2.18 | 50.1 | 15.2 | 3.2 |
| | | 20–30 | 2.14 | 49.2 | 17.1 | 2.8 |
| | | 40–50 | 1.17 | 49.0 | 20.8 | 2.3 |
| | | 50–60 | 1.62 | 39.4 | 19.7 | 2.0 |
| 7 | *Calcareous chernozem* | 0–20 | 2.46 | 44.0 | 20.8 | 2.1 |
| | | 30–40 | 2.38 | 45.2 | 19.5 | 2.3 |
| | | 50–60 | 1.71 | 42.7 | 19.3 | 2.2 |
| 8 | *Calcareous chernozem* | 0–20 | 1.55 | 38.7 | 25.4 | 1.5 |
| | | 30–40 | 1.49 | 41.5 | 21.5 | 1.9 |
| | | 50–60 | 1.08 | 42.6 | 23.3 | 1.8 |
| 9 | *Xerophyte-wooded chernozem* | 0–10 | 5.53 | 47.1 | 25.9 | 1.8 |
| | | 30–40 | 2.66 | 50.1 | 20.3 | 2.4 |
| | | 50–60 | 2.14 | 46.3 | 21.6 | 2.1 |
| 10 | *Xerophyte-wooded chernozem* | 0–10 | 2.78 | 48.1 | 22.3 | 2.1 |
| | | 30–40 | 2.48 | 50.4 | 20.8 | 2.4 |
| | | 50–60 | 1.28 | 48.8 | 22.2 | 2.2 |
| 11 | Compact *Leached chernozem* | 0–10 | 3.45 | 50.9 | 17.2 | 2.9 |
| | | 20–30 | 2.83 | 55.9 | 12.7 | 4.5 |
| | | 40–50 | 2.39 | 57.8 | 10.8 | 5.3 |
| | | 60–70 | 2.27 | 58.1 | 9.8 | 5.9 |
| | | 90–100 | 1.71 | 52.3 | 7.9 | 6.6 |
| 12 | Compact *Typical chernozem* | 0–20 | 2.31 | 48.4 | 19.2 | 2.5 |
| | | 20–30 | 2.37 | 49.4 | 13.0 | 3.8 |
| | | 40–50 | 1.67 | 54.8 | 11.8 | 4.6 |
| | | 60–70 | 1.60 | 50.8 | 8.5 | 5.9 |
| 13 | *Grey forest* | 0–10 | 4.22 | 30.6 | 29.6 | 1.0 |
| | | 20–30 | 1.77 | 39.6 | 29.3 | 1.3 |
| | | 50–60 | 0.76 | 46.2 | 25.2 | 1.8 |
| 14 | *Dark grey forest* | 0–10 | 5.19 | 35.8 | 25.7 | 1.8 |
| | | 20–30 | 2.80 | 38.1 | 19.6 | 1.9 |
| | | 50–60 | 1.48 | 38.5 | 22.2 | 1.7 |
| 15 | *Brown forest* | 3–13 | 4.19 | 25.2 | 21.2 | 1.2 |
| | | 20–30 | 0.84 | 27.0 | 28.9 | 0.9 |
| | | 43–53 | 0.67 | 27.2 | 28.5 | 0.9 |

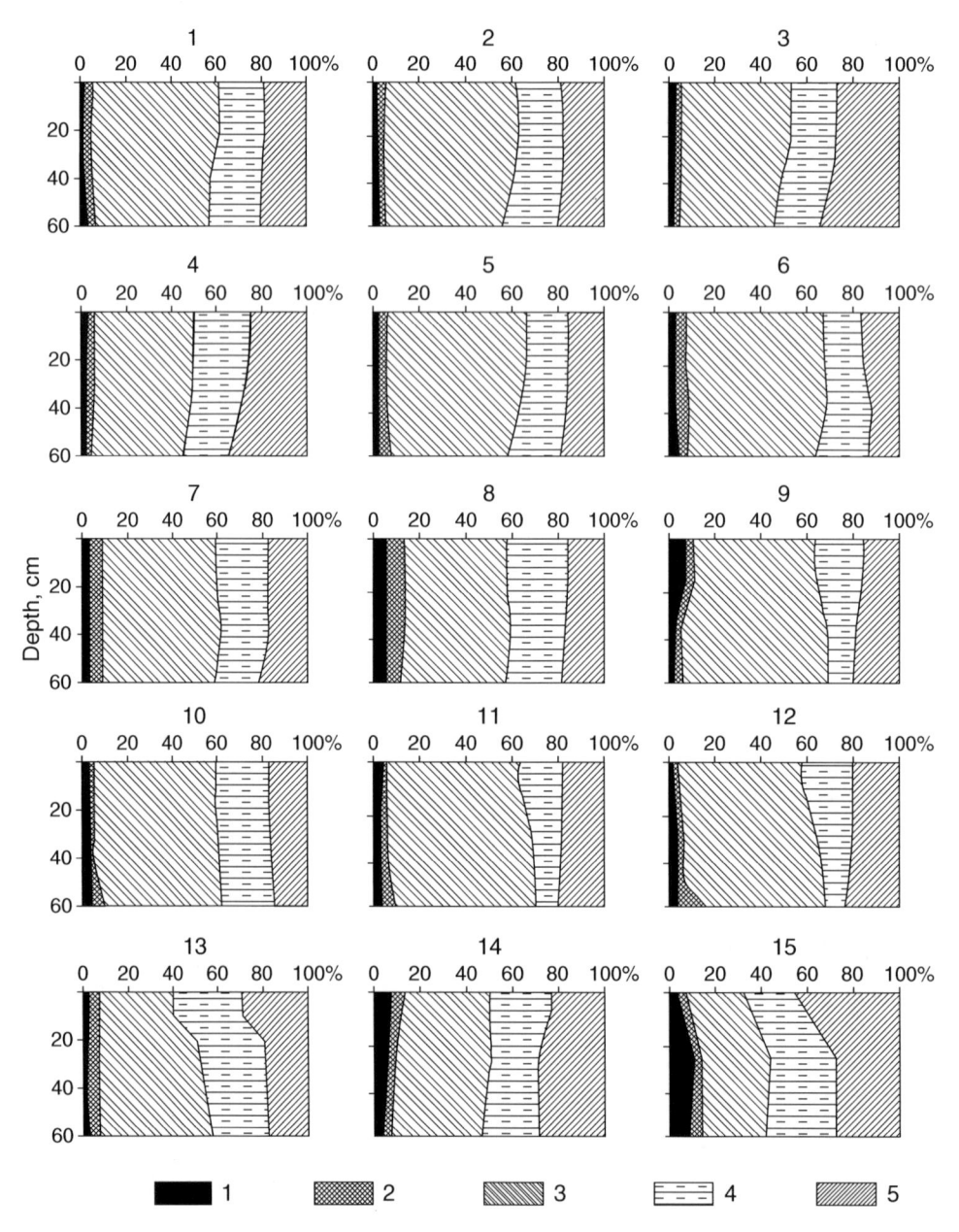

**Fig. 2** Composition of humus in Moldavian soils (percentage of soil organic carbon). *1.* bitumen, *2.* recalcitrant organic matter, *3.* humic acid, *4.* fulvic acid, *5.* humin (see Table 8 for the names of the soils)

carbonates. Figure 3 compares the soil organic carbon and carbonate-carbon profiles of *Xerophyte-wooded chernozem, Typical chernozem, Dark grey forest soils* and *Meadow soils*. In chernozem, carbonates inherited from the parent material resist depletion for a long time; leaching is counteracted by the capillary rise of groundwater saturated with calcium bicarbonate (Krupenikov 2000, Scerbacov and Vassenev 2000).

Considering the present trends in the humus and nutrient balance summarized in Table 1, we have to conclude the following:

- Land use policies in Moldova, across eastern Europe and beyond, need urgent reforms to combat the destruction of the soils and to rebuild them
- In this context, the agricultural system should be refashioned, root and branch, along ecological lines

**Fig. 3** Soil organic carbon and carbonate carbon profiles of Moldavian soils. Soil types profile number: *Calcareous chernozem* 241, 278; *Common chernozem* 303, 61, 506; *Xerophyte-wooded chernozem* 231; *Typical chernozem* 27, 21, 295; *Leached chernozem* 701, 400, 6; *Podzolized chernozem* 254; compact – 1; *Southern chernozem* – 20; *Dark-grey forest* 410; *Meadow* 23864, 25764

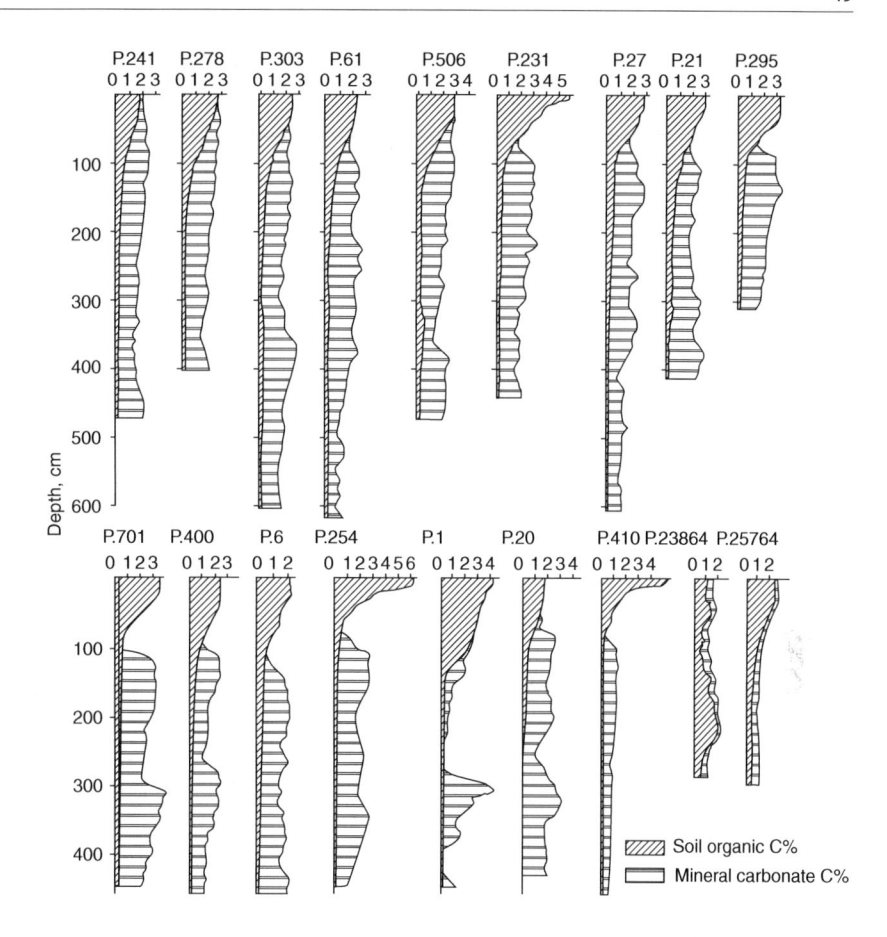

that adopt crop rotations and methods of cultivation that are productive but, at the same time, do not damage the soil

- Optimal imports and application of nitrogen and phosphate fertilizers should be provided – without any excess – not only to ensure adequate farm production but also to enable the renewal of humus reserves

# References

Andries S and C Zagorcea 2002 Soil fertility and agrochemistry service in agriculture. *Bulletin of the Academy of Sciences of Moldova: Biological, Chemical and Agricultural Sciences* 2, 42–44 (Romanian)

Batjes NH 1996 Total carbon and nitrogen in the soils of the world. *European Journal of Soil Science* 41, 2, 151–163

Bel'chikova NP 1954 Determination of soil humus by Tyurin's method. 35–42 in *Agrochemical methods of soil analysis 2nd edition*. Russian Academy of Sciences, Moscow (Russian)

M Bertholet and G André 1892 *Ann. Chim. Phys.* ser. 6, 25, 364

Brady NC and RR Weil 1996 *The nature and properties of soils*. Prentice Hall, Upper Saddle River, NJ

Burlacu I 2000 *Agrochemical service for agriculture in the Republic of Moldova*. Pontos, Chisinau, 229p (Romanian)

Cerbari V 2000 *Information system on the quality of soil cover in the Republic of Moldova (The bank of data)*. Pontos, Chisinau, 85p (Romanian)

Dokuchaev VV 1883 *Russian chernozem*. Independent Society for Economics, St Petersburg (Russian, English translation 1967 Israeli Program for Scientific Translations, Jerusalem)

Ellert BH and JR Bettany 1995 Calculation of organic matter and nutrients stored in soils under contrasting management regimes. *Canadian Journal of Soil Science* 75, 529–538

Feller C, LJM Thuries, RJ Manlay, P Robin, and E Frossard 2003 "The principles of rational agriculture" by Albrecht Daniel Thaer (1752–1828). An approach to the sustainability of cropping systems at the beginning of the 19th century. *Journal of Plant Nutrition and Soil Science* 166, 687–698

Ganenco VP 1991 *Soil humus in Moldova and transformation under the influence of fertilization*. Stiinta, Chisinau, 131p (Russian)

Glazovskaya MA 1996 Role and functions of the pedosphere in the geochemical carbon cycle. *Pochvovedenie* 2, 174–186 (Russian)

Grati VP 1977 *Forest soils of Moldova and their optimal utilization*. Stiinta, Chisinau, 128p (Russian)

GVGK 1978 *Atlas of SSR Moldavia*. GVGK, Chisinau, pp 50–56 (Russian)

IGPCC 2010 – The Carbon cycle. http://www.climatescience.gov/Library/stratplan2003/final/annexcfigure7-1.htm

Jobbagy EG and RB Jackson 2000 The vertical distribution of soil organic carbon and its relation to climate and vegetation. *Ecological Applications* 10, 2, 423–436

Korduneanu PN 1985 *The biological cycling of the major nutrients for agricultural crops 2nd edition*. Stiinta, Chisinau, 269p (Russian)

Kovda VA (editor) 1983 *Russian chernozem 100 years after Dokuchaev*. Nauka, Moscow, 308p (Russian)

Krupenikov IA 1967 *Chernozem of Moldova*. Cartea Moldoveneasca, Chisinau, 427p (Russian)

Krupenikov IA (editor) 1981 *Statistical parameters for soil composition and properties in Moldova Vol.2*. Stiinta, Chisinau, 253p (Russian)

Krupenikov IA 1987 Chernozem of Europe and Siberia: similarities and differences. *Pochvovedenie* 11, 19–23 (Russian)

Krupenikov IA 2000 Human pressures driving the destruction of chernozem as a soil type. 303–313 in AP Scerbarov, II Vassenev (editors) *Anthropogenic evolution of chernozem soils*. Voronej State University, Voronej (Russian)

Liebig J 1840 *Chemistry in its application to agriculture and physiology*. (Published in Russian, Moscow, 1864, 328p)

Liebig J 1851 *Familiar letters on chemistry* (34th letter, third edition, p 519), London

Mikhailova EA, RB Bryant, II Vassenev, SJ Schwager and CJ Post 2000 Cultivation effects on soil carbon and nitrogen contents at depth in the Russian chernozem. *Soil Science Society of America Journal* 64, 738–745

Mikhailova EA and CJ Post 2006 Organic carbon stocks in the Russian chernozem. *European Journal of Soil Science* 7, 330–336

Newton JD, FA Wyatt and AL Brown 1945 Effects of cultivation and cropping on the composition of some western Canadian soils III. *Scientific Agriculture* 25, 718–737

Quéré C le, M Raupach, JG Candell, G Marland et al. 2009 Trends in the sources and sinks of carbon dioxide. *Nature Geoscience* online 17 Nov 2009, 831–836

Reinl E 1984 Changes in soil organic carbon due to agricultural land use in Alberta. MSc thesis, University of Alberta, Edmonton AB (cited by VandenByggart and others 2003)

Scerbacov AP and II Vassenev 2000 *Anthropogenic evolution of chernozem soils*. Voronej State University, Voronej, 411p (Russian)

Sorokina NP and VM Kogut 1997 The dynamics of humus content in arable chernozem and approaches to its study. *Eurasian Soil Science* 30, 2, 146–151

Thaer A 1809–1812 Grundsätze de rationellen Landwirtschaft (The foundation for reasonable agriculture) Vol.1–5. *The history of agriculture* published in 1940, Translated in Russian 1830–1835 and in English 1844. New Russian translation, Soviet Academy of Sciences, Moscow

Ursu AT (editor) 1984 *Soils of Moldova. Vol.1 Genesis, ecology, classification and systematic description of soils*. Stiinta, Chisinau, 351p (Russian)

Ursu AT (editor) 2000 *Soil degradation and desertification*. Academy of Sciences of Moldova, Chisinau, 307p (Romanian)

Ursu AT 2005 Example of monitoring the dehumification for *Typical chernozem. Pedogenic factors and processes from the temperate zone 1V (new series, University AI Cuza, Iasi)*, 29–34 (Russian)

VandenBygaart AJ, EG Gregorich and DA Angers 2003 Influence of agricultural management on soil organic carbon: A compendium and assessment of Canadian studies. *Canadian Journal of Soil Science* 83, 363–380

Viliams VR (editor) 1940 *The history of soil fertility. Vol.1. The science of soil fertility in the 19th century*. Selihozgiz, Moscow and Leningrad, 428p (Russian)

Zagorcea KL 1990 *Optimization of fertilization systems for field crop rotations*. Stiinta, Chisinau, 269p (Russian)

# The Nitrogen Riddle

**Abstract** Nitrogen is the nutrient required in the greatest amount by plants; it is a main component of proteins which are the building blocks of life. It enters the soil from atmospheric precipitation and through the activities of micro-organisms – both free living and in symbiosis with legumes. Organic forms of nitrogen make up 5% of humus and their mineralization, through the decomposition of humus, is the main source of nitrogen for plants. Under cultivation, chernozem are losing humus and nitrogen (in Moldova, about 1.1 t/ha of humus and 36 kg/ha of nitrogen are lost from the topsoil every year). Even so, under ecological agriculture that stops or at least minimizes further losses, the humus and nitrogen reserves in full-profiled chernozem remain enough to lend resilience to the soil and the farming system. Nitrogen is available to crops mainly as the nitrate ion. Nitrate is very soluble in water and not adsorbed by the ion-exchange complex so it might be easily leached from the soil. However, it is actively taken up by plant roots and micro-organisms. Surplus is rare in natural systems but there are some cases of very high nitrate content, even in the absence of nitrate fertilizers. In southern chernozem, nitrates accumulate in the subsoil because of mineralization of organic nitrogen accompanied by deep leaching in spring, especially on sown land where evaporation is less than rainfall; in years with abundant spring rains, nitrates may even be leached below 4 m; later in the season, roots do not take up the nitrate from below about 2 m. The sink may hold more than 1 t/ha of nitrate that is mostly inaccessible to crops, but long-term trials demonstrate a natural supply of nutrients: over 25 years, corn yields with NP fertilizer averaged 4.36 t/ha compared with the control 4.27 t/ha; and 6.0 t/ha with NPK fertilizer compared with the control of 4.48 t/ha. We cannot but be impressed by the fertility of the control – that is without fertilizer! The problems of ensuring continued sufficiency of nitrogen for agriculture and arresting its leakage into the environment are global issues. Research is needed urgently into (1) reducing the losses of soil organic matter and rates of fertilizer application; (2) use of deep-rooting crops to intercept deeply leached nitrates; (3) mechanisms for drawing upon deep nitrate stocks.

**Keywords** Plant nutrients · Humus · Nitrate accumulation · Nitrate leaching

## Nitrogen in Soils and Agriculture

Saltpetre or nitre (potassium nitrate) has been known since at least the thirteenth century as the essential ingredient of gunpowder. Its agricultural significance was established in the seventeenth century by the German chemist, Glauber, who prepared nitric acid from nitre obtained in the traditional way from the scrapings of cattle byres. He argued that, if there was nitre in cattle droppings and urine, it must be a part of the plants that the cattle had eaten (Glauber 1656). By adding nitre to the soil, he achieved enormous increases in crops and he advanced the hypothesis that nitre was the "irreplaceable factor for vegetation": the value of soil and manures (he mentions dung, hair, feathers, horn, bone and rags) was entirely due to nitre; other needs coming from what he called "life power". This large idea was in opposition to Van Helmont who had previously argued that water was the *principle* of vegetation. Viliams (1940) dismissed it all as "alchemical ravings" – echoes of the quest for the philosopher's stone that would turn base metals into gold.

We now know that nitrogen is the mineral nutrient required in the greatest amount by crops; it is a main component of proteins, the building blocks of life. Nitrogen enters the soil from atmospheric precipitation and through the activities of both free-living micro-organisms and, more importantly, those living symbiotically in the roots of legumes and a few other species. A suite of other nutrients is necessary for the growth of plants, but the outstanding role of nitrogen fertilizer in agricultural production is not in doubt; modern agriculture is addicted to it. In the southern chernozem, nitrates also play an enigmatic role that is yet to be satisfactorily explained.

Thirty-five years ago, the stocks of humus in the 0–20, 0–50 and 0–100 cm layers, and for the 0–50 cm layer in respect of nitrogen, were for *Calcareous chernozem* – 70, 165 and 280 t/ha of humus and 12 t/ha of nitrogen; for *Common chernozem* – 90, 200 and 330 t/ha humus and 14 t/ha nitrogen; for *Xerophyte-wooded chernozem* – 115, 240 and 370 t/ha humus and 17 t/ha nitrogen; for *Typical chernozem* – 115, 250 and 400 t/ha humus and 17 t/ha nitrogen; for *Leached chernozem* – 112, 240 and 370 t/ha humus and 16 t/ha nitrogen; and for *Podzolized chernozem* – 100, 250 and 360 t/ha humus and 14 t/ha nitrogen (Krupenikov 1967). Summing the annual humus loss from the top 50 cm of soil since that time, all chernozem have lost about 40 t/ha of humus and 2 t/ha of nitrogen; present day annual losses are of the order of 1.1 t/ha of humus and 36 kg/ha of nitrogen (Table 1 of chapter "Humus – Guardian of Fertility and Global Carbon Sink"). Even so, the humus and nitrogen reserves in full-profiled chernozem remain enough to lend resilience to the soil and farming system under ecological agriculture that stops or, at least, minimizes further losses.

Nitrogen is available to plants mainly as the nitrate ion ($NO_3^-$). Nitrate is very soluble in water and is not adsorbed by the ion-exchange complex, so that it may readily be leached; however, it is actively taken up by plant roots and micro-organisms. Surplus is rare in natural systems (Zveagintev 1987) but there are some cases of very high nitrate content, even in the absence of nitrate fertilizers. Kudreavteva's monumental work *Nitrate in the soil, its accumulation by means of cultivation* (Kudreavteva 1927) demonstrated that the nitrate can accumulate in the soil through the decomposition of humus and that this process can be accelerated substantially by cultivation; but his experiments were confined to the top metre – and the loss of the humus itself was ignored; D.N. Prianishnicov, in his 1925 book *Chemicalization of Agriculture* (Prianishnikov 1965a, b), also erred in thinking that chernozem needs only phosphate fertilizer and not nitrogen. Nowadays, fertilizer practice in Moldova still follows Prianishnicov's method – but without phosphate.

In the 1950s, when the authors and Dr Antonina Rodina were mapping the soils of the country around Cahul, we assisted N. Vainberg with choosing typical areas for his fertilizer trials. He observed that, on *Calcareous chernozem*, corn did not respond to nitrogen fertilizer. Some years later, we investigated the soils to a depth of 4 m, measuring nitrate in fresh samples from eight soil profiles; in some soils, nitrate was present throughout the depth of the cuttings. There are some notes about this in the literature (Sapphoev 1958) but, following Kudreavteva and others, the prevailing view in respect of the chernozem was that nitrate never migrated deeper than 60–70 cm. However, most investigations had been made in areas with a more severe climate where the release of nitrate from humus through the activity of soil micro-organisms, which we may call nitrification, only begins late in the spring.

Figure 1, from an old source, depicts nitrates in unfertilized soils: in forest soils, nitrate is scarce and found only to a depth of 70 cm; similarly in the chernozem of northern Moldova. It is a different picture in the south: nitrate is abundant throughout the profile to a depth of 4 m, particularly in *Calcareous chernozem* – which explains Vainerg's unluckiness with the introduction of nitrogen fertilizer.

The nitrate story in *Calcareous chernozem* runs as follows. Nitrification begins when the topsoil warms and the soil is moist enough for micro-organisms to be active – in April in the south of the country. This nitrate is readily leached to deeper layers, and this process continues through early May on sown lands where evaporation is less than rainfall and the crops are still drawing on the nutrient reserves of the seed; in years of abundant spring rains, nitrates may even be leached below 4 m. Later in the season, the roots of grasses and most crops do not take up the nitrate from below about 2 m. When we presented our results to our director at the N Dimo Institute in 1962, we were given short shrift: "They're a fluke!" he retorted. "They're not trustworthy and I'm against publishing

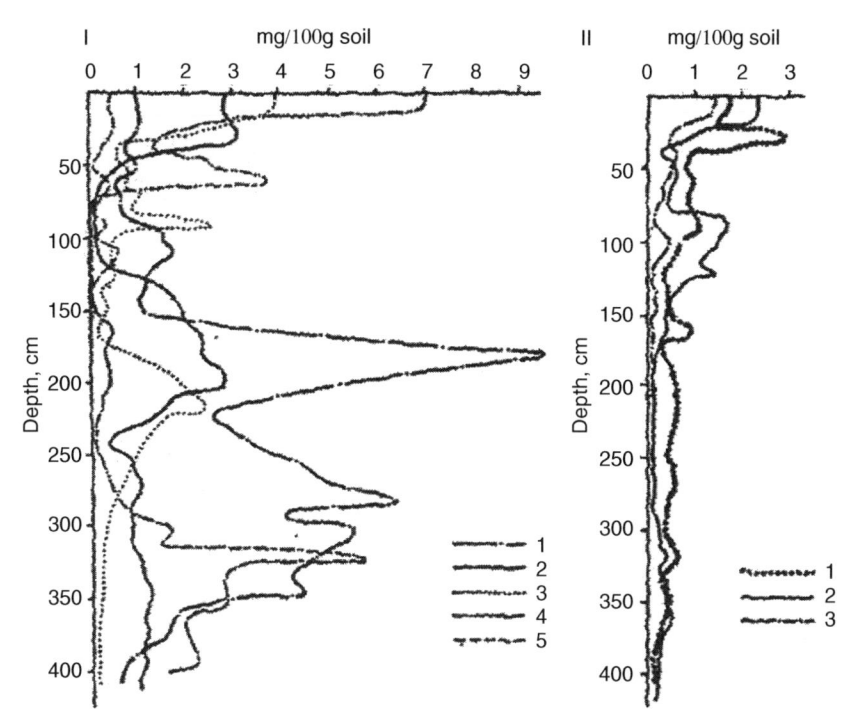

**Fig. 1** Localization of nitrates in the profiles of Moldavian soils. **I** Southern chernozem: *1. Calcareous*, ploughland, Causeni District; *2. Common*, ploughland, Anenii Noi District; *3. Common*, ploughland, Vulcanesti District; *4. Xerophyte-wooded*, ploughland, Vulcanesti District; *5. Xerophyte-wooded*, Vulcanesti District. **II** Northern chernozem: *1. Typical*, ploughland, Riscani District; *2. Typical* ploughland, Edinet District; *3. Podzolised*, forest, Edinet District

them. Now, get on with your own job!" Only some years later was Fig. 1 included in the monograph *Moldavian Chernozem* (Krupenikov 1967) – under another director.

Long-term plots on *Calcareous chernozem* at Grigorievca in Causeni District further demonstrate the natural supply of nutrients in these soils (Krupenikov 1997). Over 25 years, corn yielded on average 4.36 t/ha with NP fertilizer compared with the control 4.27 t/ha; and 6.0 t/ha with NPK fertilizer compared with the control of 4.48 t/ha. There cannot be any mistake and we cannot but be impressed by the fertility of the control, that is *without* fertilizer.

In the course of 15 years' research into current soil processes, beginning in 1967, Sinkevici (1989) confirmed the existence of a nitrate sink to a depth of 4 m, even in the absence of nitrate fertilization, and especially in the warm climate chernozem of the pre-Danubian region. We can confirm the large production of nitrate and the deep sink in *Calcareous chernozem*. The phenomenon is weaker in *Common chernozem* but, still, the sink still holds 1 t N/ha to a depth of 5 m; in *Xerophyte-wooded chernozem*, nitrate migrates to a depth of 4 m and under cultivation the sink is deeper, holding 168 kg N/ha in the 4–5 m layer; the *Typical chernozem* shows lesser nitrate migration but we see its tracks, beginning at a depth of 1 m; migration is higher in the *Podzolized chernozem*, which receive more rainfall, and the nitrate goes down to 4–5 m; deep nitrate is rare in *Grey forest soils*.

Sinkevici's results on the effects of nitrate, phosphate and potassic fertilizers are still of interest (Table 1). She explains her data in a straightforward way; nitrate can reach the groundwater and is transformed to toxic nitrite. There may have been large applications of fertilizer of which the author was unaware but, more likely, the results underline the waste of nitrate leaching to such depths – where it cannot be taken up again by plants. Unfortunately, this lesson was not learned by the agrochemical service, which took an interest in nitrogen only when the era of *big chemicalization* changed to *micro-chemicalization*

**Table 1** Nitrates in Moldavian soils, kg/ha

| Soils | Variants | Layers, depth in metres | | | | | | |
|---|---|---|---|---|---|---|---|---|
| | | 0–1 | 1–2 | 2–3 | 3–4 | 4–5 | 0–5 | >5 m |
| *Calcareous chernozem*, Cahul District | A | 1144 | 337 | 153 | 207 | 242 | 2083 | |
| *Common chernozem*, Cahul District | A | 129 | 82 | 84 | 108 | 101 | 503 | |
| *Xerophyte-wooded chernozem* (forest), Cahul District | A | 252 | 133 | 154 | 166 | 68 | 773 | |
| *Xerophyte-wooded* (ploughland), Cahul District | A | 321 | 146 | 140 | 137 | 168 | 912 | |
| *Common chernozem*, Comrat District | A | 338 | 191 | 239 | 206 | 182 | 1156 | |
| *Typical chernozem*, Glodeni District | A | 207 | 32 | 14 | 14 | 29 | 295 | 92 |
| | B | 280 | 32 | 13 | 33 | 29 | 387 | |
| *Leached chernozem*, Orhei District | A | 51 | 20 | 18 | 15 | 22 | 126 | 272 |
| | B | 81 | 98 | 44 | 76 | 104 | 398 | |
| *Leached chernozem*, Ocnita District | A | 43 | 16 | 49 | 39 | 27 | 174 | 354 |
| | B | 115 | 85 | 149 | 80 | 99 | 528 | |
| *Grey forest soil* (ploughland) Orhei District | A | 136 | 54 | 55 | 61 | 68 | 374 | 811 |
| | B | 347 | 339 | 242 | 152 | 105 | 1185 | |

Variant A is unfertilized and variant B has received fertilizer

(during 1995–2000, on average, 6 kg of nitrogen, 40 g of phosphate and 100 g calcium were applied per hectare).

Table 2 records three separate long-term experiments in which sampling to 10–20 m enables us to determine the size of the nitrate sink. On *Calcareous chernozem* at Ketros, on a terrace of the Bik river (Zagorcea 1999), nitrate was measured after 25 years of application of (I) manure at an annual dose of 19 t/ha plus 240 kg/ha phosphate fertilizer; and (II) manure at 95 t/ha + 480 kg/ha N, 360 kg/ha P and 300 kg/ha K. In the untreated plots, the amount of nitrate in the 2–10 m layer was 51 kg/ha; under treatment I, it was 1 t/ha; under treatment II it was 0.5 t/ha. This deep nitrate is inaccessible to plants. The soil is of medium loamy texture, which may increase the depth of annual wetting but only down to 2–4 m. Below this depth, according to G. Visotski, the soil water potential never exceeds the level at which water can move through this layer by gravity or capillary action; it is a *dead layer*. Applying manure in such doses in combination with nitrate fertilizer is a waste of organic and mineral nitrogen.

Comparable results obtained on a clay loam *Common chernozem* at Kirsovo, near Comrat in the south of the country (Donos 2003). Measurements 17 years after the introduction of fertilizers showed large amounts of deep nitrate: on land receiving an annual

**Table 2** Nitrate, kg/ha, in soil layers to 10–20 m

| | *Calcareous chernozem* | | | *Common chernozem* | | | *Leached chernozem* | | |
|---|---|---|---|---|---|---|---|---|---|
| Depth (m) | Control | Manure 19 t + N 240 | Manure 95 t + N 480 P 360 | Control | N 60 K 60 P 60 | N 180 K 60 P 60 | Control | N 60 K 60 P 60 | N 180 K 60 P 60 |
| 0–2 | 30 | 575 | 552 | 104 | 234 | 334 | 67 | 59 | 139 |
| 2–5 | 6 | 490 | 207 | 125 | 165 | 274 | 40 | 48 | 99 |
| 5–8 | 9 | 501 | 191 | 58 | 84 | 280 | 49 | 99 | 106 |
| 8–10 | 6 | 64 | 180 | 38 | 53 | 22 | 34 | 66 | 48 |
| 10–12 | – | – | – | 46 | 30 | 46 | 34 | 45 | 54 |
| 12–15 | – | – | – | 95 | 47 | 85 | 41 | 68 | 43 |
| 15–18 | – | – | – | 70 | 89 | 90 | 41 | 75 | 74 |
| 18–20 | – | – | – | 45 | 52 | 57 | 28 | 47 | 40 |
| 0–10 | 51 | 1630 | 1130 | 325 | 536 | 970 | 190 | 272 | 391 |
| 2–10 | 21 | 1055 | 578 | 321 | 302 | 636 | 123 | 213 | 243 |
| 0–20 | – | – | – | 471 | 709 | 1248 | 334 | 507 | 602 |
| 2–20 | – | – | – | 367 | 475 | 914 | 267 | 448 | 464 |

dose of 60 kgN/ha, the surplus nitrogen in the 2–20 m layer was 108 kg; with a dose of 180 kgN/ha, the surplus was 278 kg/ha – enough to fertilize 2–7 ha! In the *dead layer* at 5–8 m, the total quantity of nitrates under the 180 kg N/ha dose was 280 kg/ha, which is 222 kg greater than the control. How did those nitrates get there? In this instance, there is no groundwater within 20 m so no capillary fringe influencing the dead layer. Another riddle!

The *Leached chernozem* at Ivancha, in Orhei district, fertilized in the same way but sampled after 25 years (Turtureanu and Leah cited by Krupenikov 1992) shows a somewhat less pronounced accumulation of nitrate in the deep subsoil compared to the control.

## Nitrates in the Environment

The issue of nitrates for agriculture and the environment, illustrated by Fig. 2, is a global problem that becomes more acute year by year. In the case of chernozem soil management, issues that require urgent new research include the following:

- The need for nitrogen fertilizer in conventional farming practice, both for increasing yield and grain quality and for reducing the losses of soil organic matter
- Rates of nitrogen fertilizer, taking account of soil nitrogen stocks
- Mechanisms for drawing upon the whole stock of nitrate in full-profiled chernozem, for instance: the use of deep-rooting crops in the rotation to intercept deeply leached nitrates; eschewing fertilization of spring crops, when stocks of nitrate in the topsoil are already high enough
- Restoring the balance of humus by using crop rotations, ploughing-in straw and applying farmyard manure and compost.

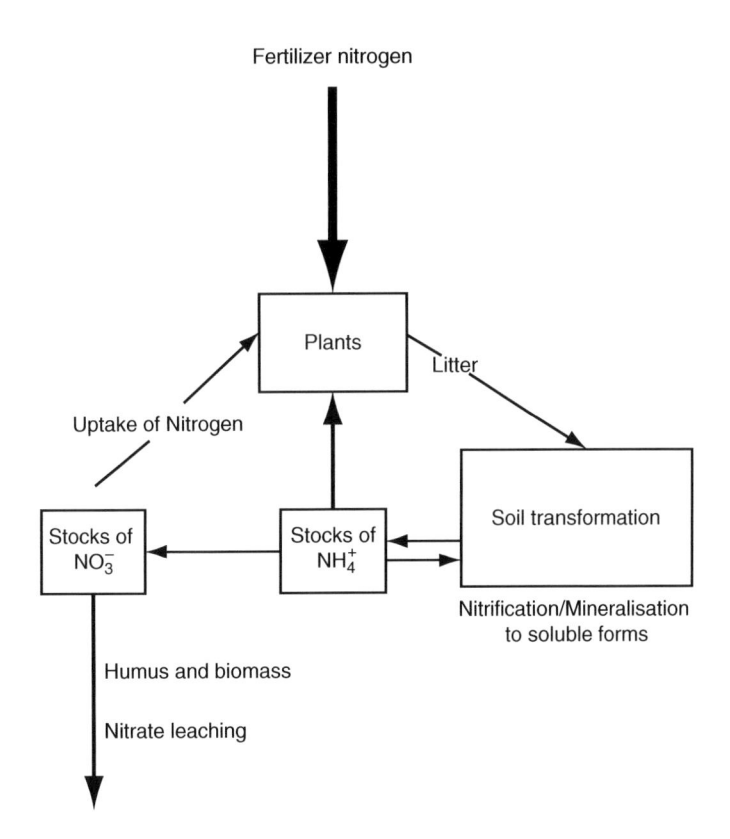

**Fig. 2** Consequences of excess nitrogen

# References

Donos A 2003 Nitrogen regime in the soils of Moldova and measures for the optimization of mineral nutrition for winter wheat. 225–227 in SV Andries (editor) *Proceedings of international conference on 50th anniversary of Institute for Soil Science and Agrochemistry: soil – one of the main problems for the 21st century, August 7 2003*. Pontos, Chisinau (Romanian)

Glauber JR 1656 *Des Teutschladis Wohlfart (erster theil), das dritte Capittel. De concentratione Vegitabitium*. Miraculum Mundi, Amsterdam

Krupenikov IA 1967 *Chernozem of Moldova*. Cartea Moldoveneasca, Chisinau, 427p (Russian)

Krupenikov IA 1992 *Soil cover of Moldova*. Stiinta, Chisinau, 264p (Russian)

Krupenikov IA 1997 Desertification of the territories between the rivers Prut and Dniester in connection with the peculiarities of soil cover. *Agriculture in Moldova* 1, 10–13 (Russian)

Kudreavteva AA 1927 *Nitrates in soils; accumulation of nitrates by soil tillage*. Edited by AG Doiarenco, M&S Subasnicov, Moscow, 317p (Russian)

Prianishnikov DN 1965a *Selected works: agrochemistry, Vol.3*. Kolos, Moscow, 767p (Russian)

Prianishnikov DN 1965b *Selected works: general issues on agriculture and Chemicalization, Vol.3*. Kolos, Moscow, 639p (Russian)

Sapphoev SP 1958 On the vertical migration of nitrates in chernozem from the Stavropol Plateau. *Pochvovedenie* 6, 66–72 (Russian)

Sinkevici ZA 1989 *Modern processes in chernozemic soils of Moldova*. Stiinta, Chisinau, 215p (Russian)

Viliams VR 1940 *The History of researches on soil organic matter*, Academy of Sciences of the USSR, Moscow (Russian)

Zagorcea C 1999 Evolution of turnover and balance of nutrients in agro ecosystems during the last century in the Republic of Moldova. in *Land and Water Resources on 125th anniversary of NK Dimo. Vol.2*. Research Institute of Soil Science and Agrochemistry, Chisinau (Romanian)

Zveagintev DG 1987 *Soil and microorganisms*. MV Lomonosov State University, Moscow, 256p (Russian)

# Phosphorus and Sulphur Budgets

**Abstract** Phosphorus is the major plant nutrient in shortest supply in all agricultural soils although chernozem are relatively well supplied. Most authorities on soil conditions and plant growth have insisted on the need for full return to the soil of phosphate taken up by crops but this requirement has been neglected. In chernozem, substantial reserves of phosphorus have been built by biological pumping of phosphorus from the mineral to the organic form. The accumulation of organic phosphate is greatest in *Leached* and *Typical chernozem*, and least in *Calcareous chernozem*. With present restrictions on imports of phosphorus and impending global shortage of phosphate fertilizer, the question is how to mobilize soil phosphorus reserves to supply crops. Sulphur is the fourth major plant nutrient in terms of the amount required by crops; legumes and brassicas take up as much sulphur as phosphorus. Generally, chernozem soils are well provided with sulphur: in the dry lands there is gypsum in the soil and dissolved sulphate in the groundwater; and everywhere receives sulphate in rain and as dry deposition, especially from the burning of oil and brown coal.

**Keywords** Phosphorus/phosphate · Sulphur/sulphate · Nutrient cycling · Global shortage · Eutrophication · Biological pumping

## Phosphorus

Phosphate was known to Arab chemists in the twelfth century but elemental phosphorus was first isolated in 1699 by the German, Brand, who called it "cold fire". The name *phosphorus* comes from the Latin, meaning *radiant*, which is appropriate for its role in living organisms; life cannot exist without it. Plants contain 230–350 mg of phosphorus per 100 g dry matter; sea animals contain 400–1800 mg and land animals 1700–4400 mg with the largest amounts in bone and brain; our own daily need is 1–1.2 g. Multiplied by the numbers of individuals, these are colossal requirements.

Liebig, in the eighteenth century, emphasized the importance of phosphorus and insisted on its "full return" to agricultural soils. Even in the early years of mineral fertilizer use in the USSR, in the 1920s, D. Prianishnicov pointed out that, of all the fertilizers, chernozem need phosphate most of all (Prianishnikov 1965b). These appeals have remained unheeded; the soil phosphorus problem has been aggravated and the natural resources of available phosphate have been recklessly consumed. On the other hand, over-application of fertilizer or organic sources such as manure and sewage can result in high levels of mobile phosphate in soils. If these soils are close to surface waters, this phosphate may be transferred as soluble phosphate or a phosphate adsorbed on eroded soil particles, which leads to eutrophication of waters – where excess nutrients accumulate stimulating the growth of algae. These, in turn, decompose – using up all the oxygen in the water and releasing toxins.

Cropped soils inevitably lose carbon, through the decomposition of humus, and nitrogen that is taken up by crops and carried off-site. However, on the global scale, there is no shortage of these elements; the reserves in the atmosphere are infinite and there are well-proven biological methods to return them to the soil – this is one of the main tasks of ecological agriculture. In chernozem, the reserves of potassium and calcium in soils and rocks are enough to last for a long time but phosphate is another story. Phosphate is required as a plant nutrient in the same order of magnitude as nitrogen and potassium but there are no additional sources in Moldova – and accessible reserves worldwide are close to exhaustion. Within 25 years, its use in agriculture will be not just expensive but practically impossible; it has to be recycled within the farming system.

According to Prianishnikov's calculations, a 1.5–2 t/ha corn crop carries away from the soil 20–25 kg of phosphorus, 35 kg of potassium and 40–50 kg of nitrogen. Until the 1940s, regular crop yields in Bessarabia were winter wheat 0.8–0.9 t/ha; barley 0.8 t/ha; maize 0.92 t/ha (Moghileanski 1913). If we make allowance also for losses in straw, during the first 100–120 years of intensive cropping, about 1.5 t of phosphorus was carried off from each hectare of cropland, though less where fallow was practised – mainly in the south of the country.

To compare this with the stock of phosphorus in the topsoil: Chiricov (1956), generalizing data for the Soviet Union, estimated that the plough layer contains from 1 to 6 t/ha of phosphorus. For Moldavian chernozem, the amount of phosphorus in the upper 20 cm is 5 t/ha in *Calcareous* and *Xerophyte-wooded chernozem*, 3–7 t/ha in *Common chernozem*, 5 t/ha in *Typical chernozem* and 4 t/ha in *Podzolized chernozem*. The average for the southern chernozem is 4.7 t/ha and for the northern chernozem 4.2 t/ha (Table 1) and Romanian chernozem contain from 3.5 to 4.5 t/ha of phosphorus, close to the Moldavian value.

The various forms of phosphorus in soil differ in their mobility and availability to the plants. In Fig. 1, we refer to Chiricov's five fractions.

*Soluble phosphate* (I) is scarce, only 1.1–1.6% of total phosphorus (in Fig. 1 it is combined with fraction II). *Somewhat mobile mineral phosphates* (II and partly III) make up about 20% of the total soil phosphorus (and 40–60% of the total in the parent material); they are most abundant in *Calcareous chernozem*, considerably less in *Typical* and *Leached chernozem*. In *Typical* and *Leached chernozem*, the main plant-available reserve is held as *organic phosphate*. The biological pumping of phosphorus from the mineral form to the organic form is shown by the ratio of organic to mineral phosphate; the accumulation of organic phosphate can be seen to the greatest depth in *Leached* and *Typical chernozem*, and least in *Calcareous chernozem*.

The values for the various phosphate fractions in the plough layer of Moldavian chernozem (Table 2) refer to the 1960s (Krupenikov 1987). Measurements were repeated 10 years later by Podari (1974) but, unfortunately, only for the plough layer (Table 3) so that we cannot assess the behaviour of the phosphate fractions throughout the soil profile.

In spite of over-exploitation of the natural phosphate store, the chernozem still have a comfortable reserve of *total* phosphorus and a relative accumulation of phosphorus in the soil profile compared within the parent material – with the reservation that most of our data refer to some decades ago. Nowadays, in connection with the drastic reduction of imports of phosphate, the question is how to mobilize soil phosphorus reserves to supply crops.

Tiganiuc (1990) argued that, for the soils poorly provided with mobile phosphorus, a regular dose of 60 kg/ha will gradually bring them to the optimal content of 140 kg/ha – but this is far from reality. From the 1970s to 1990s, the agricultural chemistry service in Moldova periodically mapped topsoil humus content and mobile forms of phosphorus, potassium and some micro-elements across some 1.6 million hectares. Table 4 after Ion Burlacu (2000) summarizes this extensive material. By 1990, in spite of the intensified application of fertilizers, almost one-third of the land was poorly provided with phosphorus, one-third moderately provided and a little more than a third well provided. There are no up-to-date maps but, undoubtedly, there is a lot of poorly provided arable land and the area of well-provided lands is much reduced. This is the most acute agronomic problem for Moldavian agriculture, and a serious issue for all farmers on the black earth.

**Table 1** Phosphorus reserves (total of Chiricov's fractions) in Moldavian chernozem, mg/100 g of soil

| Depth (cm) | n | Max | Min | Mean | CoV, % | $K_a$ |
|---|---|---|---|---|---|---|
| 0–20 | 42 | 300 | 80 | 171 | 31 | 1.33 |
| 40–60 | 32 | 270 | 40 | 150 | 34 | 1.17 |
| 90–120 | 28 | 260 | 30 | 136 | 45 | 1.05 |
| 160–200 | 28 | 230 | 30 | 120 | 42 | 1.0 |

$K_a$: Accumulation coefficient – ratio of P in soil to P in parent material.

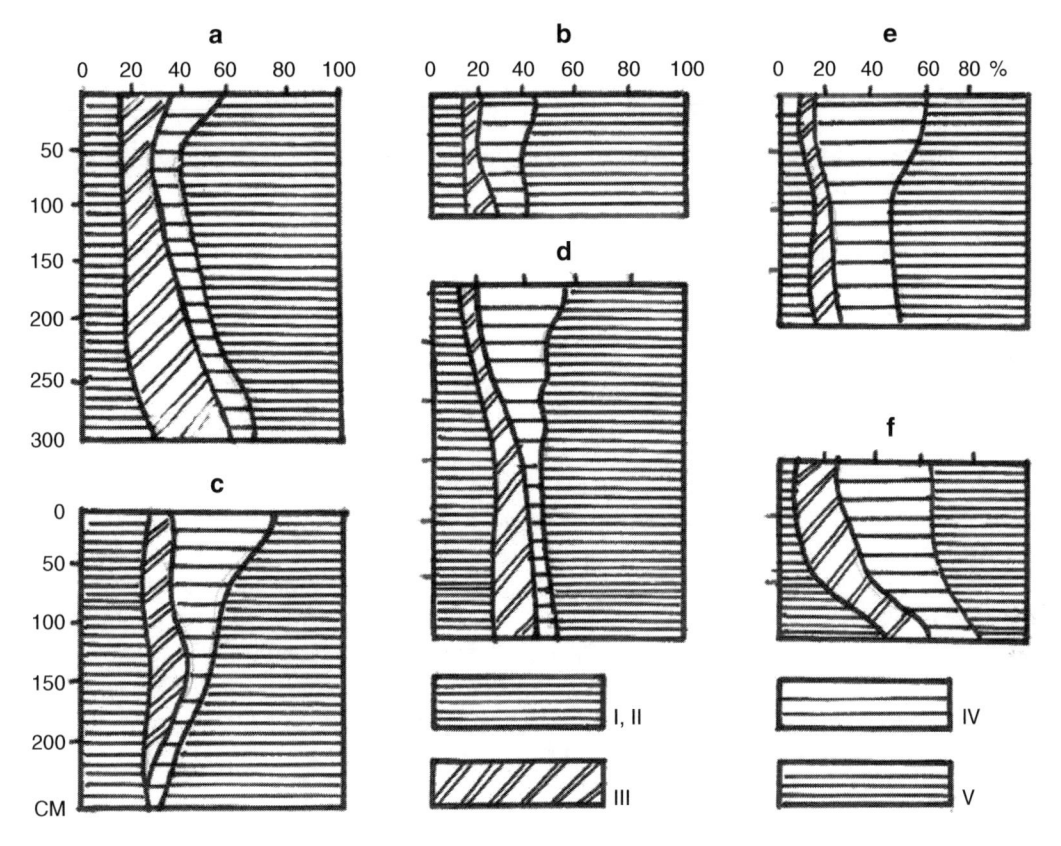

**Fig. 1** Phosphate fractions and profile distribution in chernozem, as a percentage of the total phosphate content: (**a**) *Calcareous*, (**b**) *Xerophyte-wooded*, (**c**) *Common*, (**d**) *Typical*, (**e**) *Leached*, (**f**) merged data for all chernozem; fraction I is water soluble, fractions II and III mineral phosphates, fractions IV and V organic phosphates [fraction IV is organic phosphorus soluble in 3 N ammonium hydroxide (nucleins, nucleoproteins, humic acid fractions), fraction V is not soluble in ammonium hydroxide]

**Table 2** Phosphate fractions in the plough layer of Moldavian chernozem, mg/100 g of soil

| Chernozem sub-type | Total $P_2O_5$ | Phosphorus fractions | | | | |
|---|---|---|---|---|---|---|
| | | I | II | III | IV | V |
| *Calcareous* | 140.6 | 2.0 | 26.7 | 19.4 | 62.0 | 30.5 |
| *Common* | 145.0 | 1.4 | 25.5 | 21.1 | 71.0 | 26.0 |
| *Typical* | 157.5 | 1.7 | 21.2 | 12.6 | 70.0 | 52.0 |
| *Leached* | 165.2 | 2.4 | 13.8 | 12.8 | 63.5 | 63.7 |

**Table 3** Reserves of the organic and mineral phosphorus in the 0–50 cm layer (Podari 1974)

| Chernozem sub-type | n | Organic P (t/ha) | Mineral P (t/ha) | Organic: mineral ratio |
|---|---|---|---|---|
| *Calcareous* | 241 | 2.1 | 10.9 | 0.19 |
| *Xerophyte-wooded* | 231 | 2.5 | 8.9 | 0.29 |
| *Common* | 272 | 2.9 | 4.1 | 0.70 |
| *Typical* | 295 | 3.3 | 4.7 | 0.70 |
| *Leached* | 701 | 3.7 | 2.3 | 1.60 |
| *Podzolized* | 8 | 3.6 | 3.4 | 1.06 |

**Table 4** Provision of mobile phosphorus in arable soils in Moldova, 1971–1990

| Period | Area evaluated (ha) | Adequacy of P provision, % of surveyed area | | |
|---|---|---|---|---|
| | | Low | Average | High |
| 1971–1979 | 1,712,603 | 67.7 | 21.3 | 11.9 |
| 1980–1985 | 1,674,644 | 49.2 | 27.4 | 23.4 |
| 1986–1990 | 1,592,014 | 30.8 | 33.7 | 35.5 |

## Sulphur

Sulphur is mentioned by Homer and in the Bible; it was a constituent of the terrifying *Greek fire* and, in religious rites, sulphur is burned to drive out evil. It is the fourth major plant nutrient in terms of the amount required by crops; legumes and brassicas take up as much sulphur as phosphorus. Its content in terms of dry matter is 1.2% in marine plants, 0.3% in land plants, 0.5–2% in marine animals and 0.5% in land animals.

The total amount of sulphur near the Earth's surface and cycling between the atmosphere, biosphere, hydrosphere and lithosphere is estimated to be $10^{14}$ t (Schlesinger 1997). Due to rapid oxidation and sulphate deposition, the atmosphere represents only a small stock of sulphur but it provides the main flux between reservoirs (Andrae and Jaeschke 1992). The biosphere also holds a relatively small stock but it drives the cycling of sulphur by mediating its oxidation and reduction (Giblin and Wieder 1992); most sulphur is held as dissolved sulphate in the oceans ($1.28 \times 10^{13}$ t), and in soils and rocks as gypsum ($2.47 \times 10^{13}$ t) and pyrite ($4.97 \times 10^{13}$ t).

Figure 2 depicts the cycling of sulphur through plant and soil. Sulphur, as a plant nutrient, has been somewhat neglected by research and there is a little information about sulphur in the Moldavian chernozem (Ursu 1984, Krupenikov 1967, Strijova 1973). Generally, Moldavian soils are well provided with sulphur: in the dry parts of the country there is gypsum in the soil and dissolved sulphate in the groundwater; and everywhere receives sulphate in rain and as dry deposition, especially from the burning of oil and brown coal. The sulphur content of the topsoil is usually 0.33–0.36%, rarely 0.4%; the range in the 40–70 cm layer is 0.28–0.56% and in the parent material at 100–200 cm, 0.39–0.52%. A. Strijova's small-scale maps of sulphur in soils and parent materials (Ursu 1984) show the topsoil sulphur content is 0.26–0.33% in the southern chernozem and 0.34–0.41 in the north, but with no clear differences between subtypes. Most parent materials contain 0.26–0.33%; in the Balti steppe 0.34–0.41%; close to Prut and Dniester 0.42–0.49%.

The last word on nutrient budgets, as they relate to agriculture, belongs to the official data on the import of mineral fertilizers (Table 5). In Moldova, there has been no application of potassium for at least 5 years; phosphate use is equivalent to only 100–200 g of phosphorus per hectare and, in 2001, the country imported 10 times less than in 1981–1985. We cannot claim that there is a full return of nutrients to the soil; the soil has been plundered!

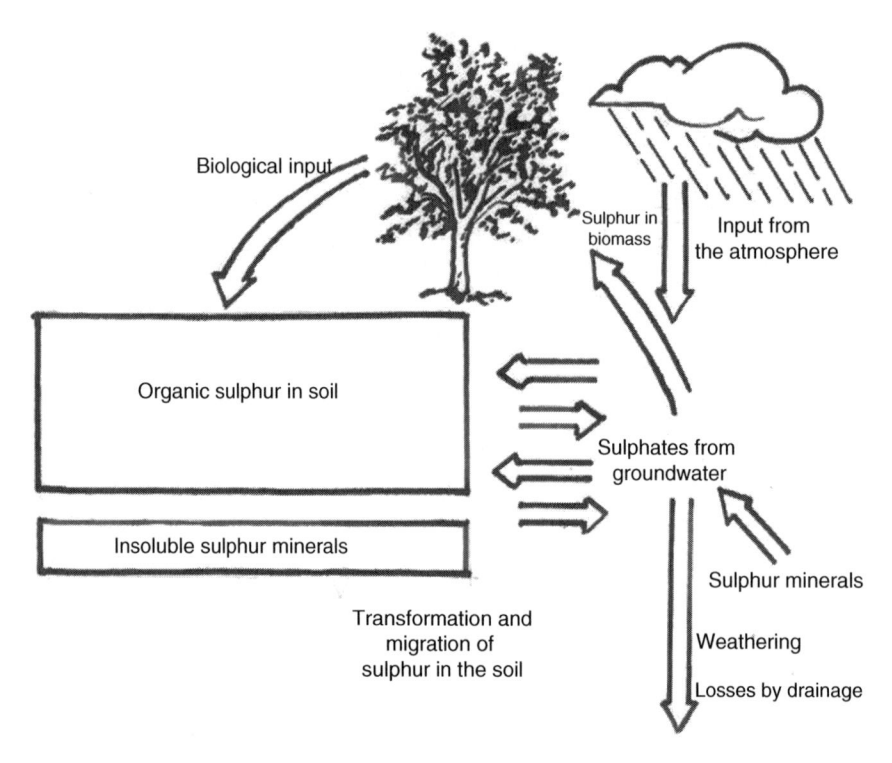

**Fig. 2** The sulphur cycle

**Table 5** Mineral fertilizer use in Moldova, 1961–2002

| Years | Thousand tonnes of active ingredient/year | | | Kg of active ingredient per 1 ha of arable crops and perennial crops | | |
|---|---|---|---|---|---|---|
| | N | $P_2O_5$ | $K_2O$ | N | $P_2O_5$ | $K_2O$ |
| 1961–1965 | 13.0 | 19.0 | 8.0 | 6.2 | 8.7 | 3.6 |
| 1966–1970 | 33.8 | 34.2 | 15.4 | 15.7 | 15.8 | 7.2 |
| 1971–1975 | 75.6 | 56.0 | 34.2 | 35.4 | 26.2 | 15.9 |
| 1976–1980 | 99.6 | 84.2 | 59.8 | 46.6 | 39.4 | 27.9 |
| 1981–1985 | 148.2 | 102.4 | 111.4 | 70.4 | 48.6 | 53.0 |
| 1986–1990 | 138.0 | 107.4 | 92.9 | 66.2 | 51.6 | 44.6 |
| 1991 | 82.7 | 75.2 | 33.5 | 53.7 | 48.7 | 21.7 |
| 1992 | 61.8 | 43.4 | 22.4 | 41.6 | 29.3 | 15.1 |
| 1993 | 20.2 | 10.6 | 5.4 | 15.1 | 7.9 | 4.0 |
| 1994 | 8.5 | 2.4 | 1.2 | 6.3 | 1.8 | 0.9 |
| 1995 | 9.6 | 1.1 | 0.5 | 7.7 | 0.9 | 0.4 |
| 1996 | 12.5 | 0.7 | 0.3 | 7.5 | 0.4 | 0.1 |
| 1997 | 9.4 | 0.5 | 0.2 | 7.5 | 0.4 | 0.1 |
| 1998 | 6.8 | 0.1 | 0 | 5.4 | 0.1 | 0 |
| 1999 | 5.7 | 0.05 | 0 | 4.5 | 0.1 | 0 |
| 2000 | 1.9 | 0.1 | 0 | 9.5 | 0.2 | 0 |
| 2001 | 13.6 | – | – | 10.9 | – | – |
| 2002 | 18.6 | – | – | 14.3 | – | – |

# References

Andrae MO and WA Jaeschke 1992 Exchange of sulphur between the biosphere and atmosphere over temperate and tropical regions. 27–66 in RW Howarth, JWB Stewart and MV Ivanov (editors) *Sulphur cycling on the continents: wetlands, terrestrial ecosystems and associated water bodies. SCOPE 48.* Wiley, Chichester

Burlacu I 2000 *Agrochemical service for agriculture in the Republic of Moldova.* Pontos, Chisinau, 229p (Romanian)

Chiricov VF 1956 *Agrochemistry of potassium and phosphorus.* Academy of Sciences of the USSR, Moscow, 276p (Russian)

Giblin AE and RK Wieder 1992 Sulphur cycling in marine and freshwater wetlands. 85–124 in RW Howarth, JWB Stewart and MV Ivanov (editors) *Sulphur cycling on the continents: wetlands, terrestrial ecosystems and associated water bodies. SCOPE 48.* Wiley, Chichester

Krupenikov IA 1967 *Chernozem of Moldova.* Cartea Moldoveneasca, Chisinau, 427p (Russian)

Krupenikov IA 1987 Chernozem of Europe and Siberia: similarities and differences. *Pochvovedenie* 11, 19–23 (Russian)

Moghileanski NK 1913 Materials on geography and statistics in Bessarabia. In *Agriculture in Bessarabia.* Chisinau (Russian)

Podari VD 1974 *The composition of phosphates in the main soils of Moldova.* Short version of dissertation, Agricultural Institute (now Agricultural University), Chisinau, 24p (Russian)

Prianishnikov DN 1965a *Selected works: agrochemistry, Vol. 3.* Kolos, Moscow, 767p (Russian)

Prianishnikov DN 1965b *Selected works: general issues on agriculture and Chemicalization, Vol. 3.* Kolos, Moscow, 639p (Russian)

Schlesinger WH 1997 *Biogeochemistry: an analysis of global change.* Academic Press, New York

Strijova TN 1973 Sulphur in soils of Moldova. *Agriculture in Moldova* 6, 26 (Russian)

Tiganiuc VD 1990 Forecast for mobile phosphate formation in chernozem soils. *Agrochemistry* 3, 12–20 (Russian)

Ursu AT (editor) 1984 *Soils of Moldova. Vol.1 Genesis, ecology, classification and systematic description of soils.* Stiinta, Chisinau, 351p (Russian)

# Life in the Soil

**Abstract** The soil is host to many and varied groups of organisms and countless individuals that make up complex food webs, decomposing the primary production of plants and recycling materials and energy. It is becoming clearer that the most effective basis for the farming of the future is likely to be husbandry and propagation of these biological systems, and we need to understand them. The crotovinas (infilled burrows) that are a conspicuous feature of chernozem are made by various large burrowing animals, but the chief architects and builders of the chernozem are invertebrates, especially earthworms. In ploughed chernozem in Moldova, the worm population ranges from 140,000 to 280,000/ha; in pristine chernozem, there are more than 1 million/ha and they make up about half of the invertebrate population. Their casts are a conspicuous feature of the soil profile. Worms augment soil porosity, aeration and water infiltration and drainage; the contributions of this unseen workforce to agriculture and the development of whole landscapes were perceived by Darwin in the nineteenth century but neglected by agricultural science till quite recently. Moving down in size but up in numbers, the total number of micro-organisms in the upper 20 cm of the chernozem is 6 or 7 million/ha. Cellulose-decomposing bacteria and actinomycetes play an essential role in the formation of humus and ammonifiers in the nitrogen cycle. Although the nitrogen-fixing Azotobacter are numerous, especially in the south, they fix only 5–6 kg of nitrogen per hectare annually and they cannot replace the nitrogen-fixing role of legumes in agriculture.

**Keywords** Soil morphology and biodiversity · Earthworms · Decomposition and mineralization · Nitrogen fixation

## Soil Biodiversity

Vernadsky saw soil as a distillation of life – surpassing the hydrosphere 10- or 20-fold and exceeding the lithosphere and atmosphere by several orders of magnitude. The disparate life in the soil includes plants, vertebrates, invertebrates and legions of micro-organisms (bacteria, cyanobacteria, fungi, actinomyces), each group commanding a whole scientific discipline in its own right; the literature on soil biology is vast, and we have used it eclectically.

In virgin steppe, the feather grass (*Stipa* sp.) roots abundantly down to 80–100 cm and some roots extend deeper than 140 cm. In arable land, the roots of corn are abundant down to 70–80 cm (Salit 1950); though rare below this depth, they nevertheless ramify the whole of the humose layer and produce 40–55 t of root mass per hectare (Barasnina 1960) – which is completely decomposed by the middle of the following year! Comparable figures of 38–49 t of root mass per hectare per year have been recorded in chernozem of the North Caucasus and the Voronej region (Stankov 1964).

From several sources, the annual production of roots appears to be around 35–50 t/ha; together with above-ground crop residues, this provides abundant food for the many and various populations of decomposers that recycle the nutrients and energy on which soil fertility depends (Sidorov and Barasnina 1966, Krupenikov 1967, Korduneanu 1978, Ursu 1986, Zagorcea 1990, Boincean 1999).

In the chernozem, large burrowing mammals like the marmot badger, suslik, blind mole rat and mole create networks of tunnels that are plugged with black earth; these are known as *crotovinas*. More than 60 years ago, one of the authors became interested in the ecology of marmots, which are one of the largest tunnelers in the virgin *Stipa* steppe in north-western Kazakhstan and display remarkable skill in distinguishing soil texture; they feed in heavy-textured chestnut soils or *Kastanozem* but dig their winter dwellings in nearby slivers of sandy soil, mapping the boundary between the two soils with great accuracy (Krupenikov and Stepanitcaia 1943). Figure 1 shows some sketches of crotovinas, with measurements in Table 1. These are, of course, individual examples; it is hard to know the general situation, but *Xerophyte-wooded* and *Typical chernozem* are well known for their abundant crotovinas, even though the soils may have been ploughed up a long time ago and large burrowing mammals are now rare or are absent.

Burrows drain and aerate the soil, and they are often sought out by roots. But the soil's chief architects and builders are invertebrates – especially earthworms. The ancients knew the soil was good if ravens followed

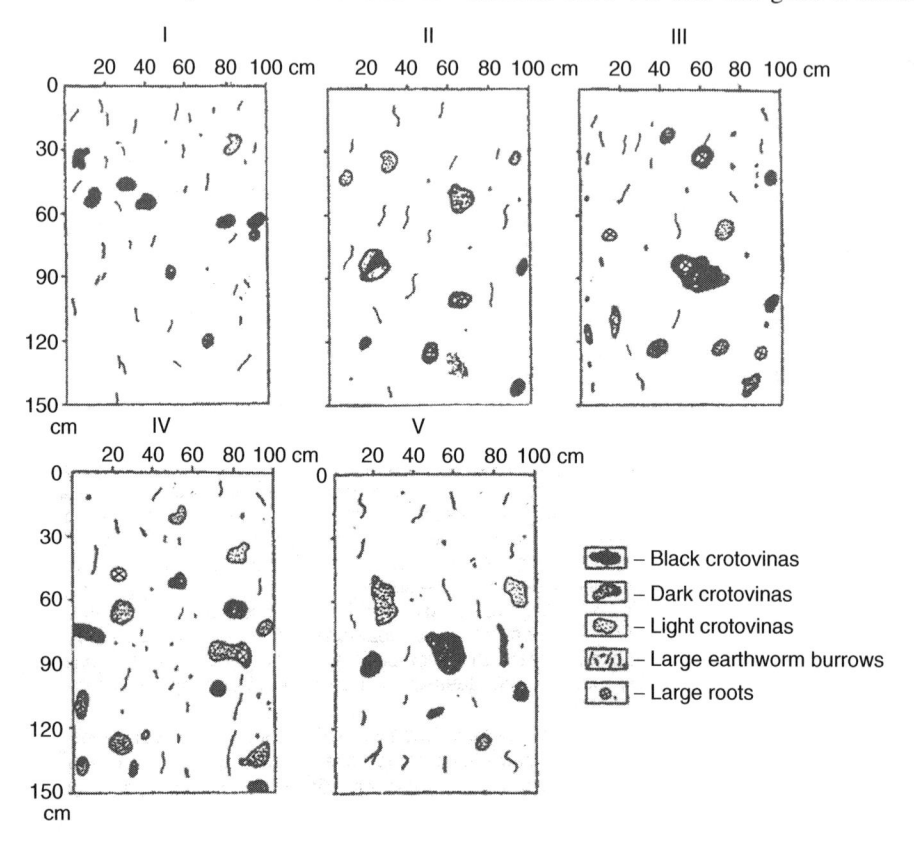

**Fig. 1** Crotovinas in chernozem

**Table 1** Data on crotovinas in chernozem

| | Chernozem sub-type, region | Number of crotovinas | | | | Average area for one crotovina (cm$^2$) | Area of crotovinas: 1 in cm$^2$; 2 as % area of pit face (15,000 cm$^2$) | | | | | | | |
|---|---|---|---|---|---|---|---|---|---|---|---|---|---|---|
| | | Black | Dark | Light | Total | | 1 | 2 | 1 | 2 | 1 | 2 | 1 | 2 |
| I | *Calcareous* Izmail, Ukraine | 6 | 3 | 1 | 10 | 44 | 316 | 2.1 | 72 | 0.5 | 56 | 0.4 | 444 | 3.0 |
| II | *Common* Vulcanesti, Moldova | 2 | 6 | 3 | 11 | 56 | 88 | 0.6 | 408 | 2.7 | 120 | 0.8 | 616 | 4.1 |
| III | *Xerophyte-wooded*, Cahul, Moldova | 3 | 7 | 1 | 11 | 78 | 470 | 3.1 | 336 | 2.2 | 56 | 0.4 | 862 | 5.7 |
| IV | *Xerophyte-wooded*, other face | 4 | 11 | 1 | 16 | 67 | 276 | 1.8 | 720 | 4.8 | 68 | 0.5 | 1064 | 7.1 |
| V | *Typical*, Riscani | 3 | 4 | 1 | 8 | 94 | 348 | 2.4 | 304 | 2.0 | 96 | 0.6 | 748 | 5.0 |

the plough to pick up the worms. The English parson, Gilbert White, observed in 1777: "Worms seem to be great promoters of vegetation which would proceed but lamely without them, by boring, perforating and loosening the soil and rendering it pervious to rains and the fibres of plants; by drawing straws and stalks of leaves and twigs into it; and most of all, by throwing up such infinite numbers of lumps of earth called worm-casts, which, being their excrement, is a fine manure for grain and grass" (White 1788).

A century later, Charles Darwin brought the role of earthworms to the attention of science in a paper *On the formation of mould* read to the Geological Society of London in 1873. Ten years later, he expanded it to a monograph (Darwin 1881) that appeared in Russian translation in 1889. He showed that worms, by ingesting and processing the soil, create the dark-coloured, friable topsoil: "All the (surface soil) over the whole country has passed many times through, and will again pass many times through, the intestinal canal of worms." And yet, 40 years of observation, experimentation and thought by the world's greatest naturalist had no influence on the agricultural science of the day – which was much exercised by the implications of Liebig's *Chemistry in its applications to agriculture and physiology*, published in 1840, and the emerging results of the well-known wheat experiments at Rothamsted; agricultural science was still a branch of chemistry.

Worms periodically expose the surface soil to the air, sift it so that no stones larger than the particles that they can swallow are left in it and intimately mix the mineral and organic material together. Their burrows augment soil porosity, aeration, and water infiltration and drainage; their casts may be clearly seen in any vertical section of chernozem; unfortunately, nobody has tried to count them as a proportion of structural aggregates but, doubtless, they contribute significantly to the water-stable aggregates that are so important for agriculture. Dimo's *Earthworms in soils of Central Asia*, published to commemorate the 50th anniversary of Darwin's first report, records between 600,000 and 720,000 earthworms in a hectare of *Dark grey soils* (comparable to *Calcareous chernozem*) under steppe; 1,300,000 earthworms per hectare in irrigated arable; and 5 million under lucerne. Dimo counted up to 15 million earthworm burrows on 1 ha of cultivated soil (Dimo 1938, 1955), which might suggest that the worm population serves as an indicator of cultivation, although most observations show that the worm population declines abruptly when grassland is ploughed up.

In ploughed chernozem in Moldova, the earthworm population ranges from 140,000 to 280,000/ha; in virgin *Xerophyte-wooded chernozem*, there are more than 1 million/ha in the topsoil alone (Table 2), and they make up about half of the invertebrate population. The foremost Russian expert on earthworms has identified 20 separate species in Moldova and counted 236 individuals in a square metre (2.36 million/ha) in *Xerophyte-wooded chernozem* (Prohina 1968); this compares favourably with 66–80 per $m^2$ in virgin chernozem on the Kursk steppe and 38 per $m^2$ on the Barabinskaya steppe in Siberia (Pereli 1972). These are not casual observations; the data are means of many measurements, and it is reasonable to deduce from them the influence of ecological conditions.

**Table 2** Mesofauna, micro-organisms and respiration in the 0–30 cm layer

| Chernozem sub-type | Soil respiration $CO_2$, volume % | Mesofauna Number/$m^2$ | | Micro-organisms % of total amount | | | | Azotobacter |
|---|---|---|---|---|---|---|---|---|
| | | Total | Earthworm | Total thousand/g | Bacteria | Actinomycete | Fungi | Number/g |
| *Calcareous*, arable | 0.8–1.4 | 38 | 14 | 6294 | 44.2 | 55.2 | 0.6 | 850 |
| *Common*, arable | 3.8–4.2 | 44 | 23 | 6645 | 47.4 | 52.0 | 0.6 | 380 |
| *Typical*, arable | 5.4–6.4 | 61 | 30 | 6000 | 57.6 | 41.6 | 0.8 | 200 |
| *Xerophyte-wooded*, arable | 5.0–7.0 | 62 | 28 | 6876 | 39.8 | 59.8 | 0.4 | 400 |
| Ditto, forest | 12.5–16.4 | 242 | 104 | 6894 | – | – | – | 0 |
| *Leached*, arable | 5.7–7.3 | | | 6200 | 62.2 | 36.6 | 1.2 | 60 |
| *Leached*, forest | 13.0–15.2 | 141 | – | 6010 | 64.4 | 34.8 | 0.8 | 0 |

Moving down in size but up in numbers, we find that the total number of micro-organisms in the upper 20 cm of the chernozem is 6 or 7 million/ha. Data collected by Mehtieev indicate that, in the northern sub-types (*Typical* and *Leached chernozem*), bacteria are most numerous; in the southern chernozem (*Calcareous, Xerophyte wooded*), the actinomyces – which are responsible for characteristic scent of freshly ploughed earth and from which several valuable antibiotics are derived. It is fascinating to note that farmers in ancient Greece would place newborn babies on the ploughed soil to give them health and vigour.

The number of micro-organisms does not indicate their activity, which may be better gauged by the soil's respiration. Nowadays, this is expressed as milligram $CO_2$ emitted per mass of soil per day, but measurements of volume % from the 1970s and 1980s[1] are still useful as a relative measure: on ploughland, soil respiration is higher in the northern than in the southern chernozem (5.4–7.3 volume % compared with 0.8–4.2%; exceptionally, 5–7% is recorded in cultivated *Xerophyte-wooded chernozem* and 12.5–16.4% in woodland; high values are also recorded for *Leached chernozem* under woodland (13–15.2%)). These data have been confirmed more recently by Marinescu (1991).

Zakharov's research on cellulose-decomposing bacteria demonstrated their essential role in the formation of humus (Zakharov 1978). Demcenco and Marinescu (1999) further demonstrated that the top 25 cm layer of full-profiled chernozem is rich in ammonifiers (in *Typical* and *Common chernozem* 5 and 5.5 million/ha, respectively; in *Leached chernozem* and *Calcareous chernozem* 3.6 and 3.2 million/ha, respectively). Although the nitrogen-fixing Azotobacter are numerous, especially in the south, they fix only 5–6 kg of nitrogen per hectare annually (Misustin 1981) and they cannot replace the nitrogen-fixing role of legumes in agriculture.

We conclude this chapter with another paradox. Speaking of the organisms living in the soil contrasts them with the non-living soil – but nearly all soil organisms eat organic matter, whether this organic matter be alive or dead, and their activities transform the soil's inert components mechanically, physically and even chemically. We may doubt Vernadsky's concept of *ecumene* or the biosphere as a framework for agricultural science – there are natural forces that do not depend on life (such as volcanism, tectonics, denudation, oceanic currents and atmospheric circulation) and, yet, determine its evolution and geography – but the most effective basis for the farming of the future is likely to be husbandry and propagation of biological systems.

## References

Barasnina LN 1960 Accumulation and decomposition of soil organic matter from perennial leguminous crops, winter wheat and maize. 237–256 in MI Sidorov (editor) *Soil tillage in Moldavia*. Cartea Moldoveneasca. Chisinau (Russian)

Boincean BP 1999 *Ecological farming in the Republic of Moldova (crop rotation and soil organic matter)*. Stiinta, Chisinau, 269p (Russian)

Darwin C 1881 *The formation of vegetable mould through the action of worms with observations of their habits*. John Murray, London, 328p

Demcenco EN and KM Marinescu 1999 The diversity of soil invertebrates and microorganisms; state, protection and perspectives. 114–131 in AT Ursu (editor) *Soil science in Moldova at the end of the second millennium*. Stiinta, Chisinau (Russian)

Dimo NA 1938 Earthworms in soils of Middle Asia. *Pochvovedenie* 4, 494–526 (Russian)

Dimo NA 1955 *Observations and researches on soil fauna*. State Publisher of Moldavia, Kishinev, 155p (Russian)

Korduneanu PN 1978 *Turnover of the major nutrients for agricultural crops*. Stiinta, Chisinau, 139p (Russian)

Krupenikov IA and SM Stepanitcaia 1943 The influence of marmot badger *(Marmotia bobac)* on soils in connection with some characters of its ecology. *Journal of Zoology* 6, 369–373 (Russian)

Krupenikov IA 1967 *Chernozem of Moldova*. Cartea Moldoveneasca, Chisinau, 427p (Russian)

Marinescu KM 1991 *Microbial ecology on ameliorated soils*. Stiinta, Chisinau, 153p (Russian)

Misustin EN 1981 *Microbial communities and their functionality in soils*. Naukova Dumka, Kiev, pp 3–13 (Russian)

Markarov BM 1970 Regarding the method for determining soil respiration of $CO_2$. *Pochvovedenie* 5

Pereli TS 1972 *Range and regularities in the distribution of earthworms in the USSR fauna*. Nauka, Academy of Sciences of the USSR, Moscow, 272p (Russian)

Prohina NA 1968 *Soil mesofauna in chernozem of Moldova*. Dissertation abstract, Academy of Sciences of Moldova, Chisinau, 20p (Russian)

Salit MS 1950 The underground part of some meadow, steppe and desert crops and ecosystems. Volume 1 Grasses and

---

[1] Determined by the now obsolete method of BM Markarov (1970), earlier proposed in 1952.

# References

semi-bush crops and ecosystems of forest (meadow) and steppe zones. 205–447 in EM Sokolov and SI Sennicova (editors) *Geobotany*. Edition 3, No. 6. Works of the Komorov Botanical Garden Institute (Russian)

Sidorov MI and LN Barasnina 1966 Accumulation and decomposition in soil of crop residues. 97–109 in Sidorov MI (editor) *Crop rotation and soil fertility in Moldavia*. Cartea Moldoveneasca, Chisinau (Russian)

Stankov NZ 1964 *The root system of field crops*. Stiinta, Chisinau, 139p (Russian)

Ursu AT (editor) 1986 *Soils of Moldova. Vol.3 Soil management, protection and improvement*. Stiinta, Chisinau, 333p (Russian)

White G 1788 Letter XXXV, May 1777. In: *The natural history of Selbourne in the County of Southampton*. White and Son, London

Zagorcea KL 1990 *Optimization of fertilization systems for field crop rotations*. Stiinta, Chisinau, 269p (Russian)

Zakharov IS 1978 *Humus formation by cellulose–decomposing organisms*. Stiinta, Chisinau, 115p (Russian)

# Soil Structure, Soil Water and Drought

**Abstract** As much as 40% of the variation in crop yields may be attributed to soil physical conditions. These are partly inherited from the parent material but others are intrinsic to the soil itself – especially the fine-granular structure that distinguishes chernozem everywhere and determines its exceptional fertility, and which is created by grass roots. Several physical properties like particle density and soil texture are stable. Others like soil structure, bulk density[1] and porosity are stable under natural vegetation but degrade under cultivation. For instance, the bulk density of the topsoil of pristine *Xerophyte-wooded chernozem* lies in the range 0.99–1.02 g/cm$^3$ whereas in ploughed chernozem it is 1–1.25 g/cm$^3$. Even so, the bulk density of the plough layer is equivalent to a total pore space of between 50 and 55% of the soil volume, which is optimal for field crops. Heavy traffic produces a compact layer (density of about 1.5 g/cm$^3$ at 50 cm) which impedes drainage and rooting. The pore space is the habitat of the soil biota, rooting space for plants, receptacle for water and nutrients in solution, and the pathway for transport of materials in solution and suspension and diffusion of gases. Most of the water held in the soil and available to plants is held in fine pores within the soil aggregates: the coarser pores between aggregates allow for infiltration of water into the soil and drainage of any surplus. Chernozem topsoils hold 16–19% by volume of plant-available water; these high values are a result of both the fine-loamy texture and the fine-granular structure. Even so, drought is a regular visitor across tracts of chernozem everywhere. In the more humid parts of Moldova, drought occurs 9 years in 29, including 3 years of severe drought; in the driest parts it occurs 12 years in 29, including 5.5 of severe drought; but there is no evidence of progressive drying. Drought and soil erosion are two sides of the same coin. When soil permeability is impaired, intense rains cause destructive runoff that washes away the soil. The less the control of surface runoff, the more the erosion and the greater the likelihood of shortage of water for crops. The farmers who come through best are those who care best for their soil – through crop rotations that achieve the best possible ground cover and by mulching bare soil. Irrigation is an effective insurance against drought, but it is costly and we should first make the best use of the rain by maintaining the soil as a free reservoir and promoting recharge by promoting deep percolation.

**Keywords** Soil physical conditions · Soil structure · Plant-available water capacity · Drought · Resilience against soil erosion

## Biological Control on Soil Physical Attributes

The attributes of soils vary systematically, both vertically down the soil profile and laterally, uphill and down dale in *catenas* and across farmers' fields. Recent research suggests that some 40% of the variation in crop yields may be attributed to soil physical conditions (Berezin and Gudima 2002). These are partly inherited from the parent material and partly intrinsic to the soil itself – especially the characteristic micro-aggregation that distinguishes chernozem everywhere

---

[1] The density of the undisturbed dry soil inclusive of its air-filled pore space.

and which determines its permeability to air, water and roots, its tilth and resistance to erosion – in short its exceptional fertility under agricultural use (see for Moldova, Krupenikov 1967; for the Ukraine, Medvedev 1988; for Russia, Scerbacov and Vassenev 2000).

From the agricultural point of view, the most important agents in the formation of the chernozem are grass roots, which are indispensable for the production of the water-stable granular structure that is the goal of farmers on any soil. Following G.V. Jacks (1954), sometime director of the world's first agricultural experimental station at Rothamsted, we cannot do better than quote at length the words of Bradfield (1937):

> Practically all soils as far as I know are found to be in a better physical condition after they have grown a heavy crop of grass or, preferably, a grass-legume mixture for a few years ... Under grass, most soils assume a granular or crumb structure, best developed in soils that are saturated with lime and which contain 3–10% organic matter. Having these amounts of lime and organic matter in the soil, however, will not ensure a good granular structure. Something else is needed. I have tried to picture to myself how these granules might be formed and why they are so important.
>
> Grass roots are so numerous that in a well-established sod they are seldom over 3–5 mm apart. These roots ramify the soil in all directions. Each root represents a centre of water removal. As water is removed from the soil in contact with the root, additional water moves toward the root by capillarity. As water is removed, the small fragment of soil between the roots shrinks and is blocked off by the roots. The pressure developed by the capillary forces, compressing the granule on all sides, reaches over 5,000 pounds per inch$^2$. As a result these granules become quite dense, their (dry bulk density) ranging from 1.8 to 3 g/cm$^2$. The total pore space inside them is small and the size of the pores is very small. Water moves into them slowly but is held firmly ... As a result, ventilation of the interior is poor. Consequently, reducing conditions frequently exist in the interior of the granules simultaneously with oxidizing conditions on the surface. This often causes a migration of substances which are more soluble in the reduced form to the surface of the granule where they are oxidized and deposited. This deposit serves as a cement and helps stabilize the granule.
>
> In forcing its way through the soil many cells are sloughed off the living root and serve as food for bacteria. Eventually, the roots die and are decomposed in situ, forming a humified, often water-resistant coating around the granule. The marked difference in colour between the surface of such granules and their interiors is evidence of this. In many respects, it seems to me these tiny granules may be compared with miniature Earths. Most of the inhabitants live near the surface where the air is better. In strongly granulated soil, practically the entire mass of clay and silt particles are clumped together in these water-stable aggregates. As a result there are two fairly sharply defined groups of pores in such soils, capillary pores within the granule and non-capillary pores between them. The non-capillary pores are relatively large. Water enters readily but is retained only at the periphery. This leaves a continuous series of connecting chambers through which air and water can readily pass. The water at the periphery is drawn into the capillary pores between the particles making up the granules. This water constitutes the most closely held reserve in the soil. Such a soil has a permeability approaching that of sandy soils combined with the storage capacity of heavier-textured soils.

Such are the structures which perennial grasses tend to develop in soils!

## Stable and Unstable Attributes

Information on the physics of Moldavian chernozem is scattered (Barasnina 1960, Krupenikov 1967, Sinkevici 1989, Ungureanu and Holmetchi 1990, Vladimir 1990, Nour 2001). Atamaniuk and others (1970) produced a short manual; an improved version is in preparation but some of the material is summarized in the first volume of *Soils of Moldova* (Ursu 1984).

Table 1 summarizes the physical properties of Moldavian chernozem; several fundamental properties like particle density, bulk density and porosity are quite constant within the limits of soil horizons so we only quote average values. The most stable physical attribute is the particle density, which is generally 2.6–2.7 g/cm$^3$ but less in the topsoil where the humus content is greater – for instance, in pristine chernozem where the humus content of the topsoil may reach 10%.

In natural conditions, protected by steppe vegetation, many other physical conditions are stable – but this stability is lost when the steppe is ploughed. This change may be seen from a comparison of the bulk density of the topsoil of pristine *Xerophyte-wooded chernozem* (0.99–1.02 g/cm$^3$) with that of ploughland (1–1.25 g/cm$^3$). Even so, the bulk density of the plough layer of the chernozem is equivalent to a total pore space of between 50 and 55% of the soil volume, which is optimal for field crops (Bondarev 2002). In the subsoil, the bulk density is between 1.3 and 1.4 g/cm$^3$, except in the 50–100 cm layer of *Leached* and *Podzolized chernozem w*here some of the pore space is plugged by clay washed in from the layers above and bulk density may exceed 1.5 g/cm$^3$.

**Table 1** Physical properties of Moldavian chernozem

| Soil sub-type and texture | Depth (cm) | Particle density (g/cm³) | | Bulk density (g/cm³) | | Porosity (%) | |
|---|---|---|---|---|---|---|---|
| | | n | Mean | n | Mean | Total | Air-filled porosity at field capacity |
| *Calcareous*, heavy loam | 0–20 | 36 | 2.64 | 36 | 1.20 | 54.5 | 20.7 |
| | 20–50 | 43 | 2.66 | 50 | 1.27 | 52.3 | 19.8 |
| | 50–100 | 66 | 2.69 | 82 | 1.33 | 50.7 | 18.5 |
| | 100–150 | 65 | 2.71 | 63 | 1.37 | 49.5 | 18.1 |
| | 150–200 | 5 | 2.69 | 13 | 1.41 | – | – |
| *Common*, light loam | 0–20 | 10 | 2.63 | 9 | 1.15 | 56.3 | 20.6 |
| | 20–50 | 14 | 2.66 | 15 | 1.28 | 51.9 | 17.3 |
| | 50–100 | 24 | 2.69 | 24 | 1.41 | 47.6 | 13.9 |
| | 100–150 | 21 | 2.72 | 21 | 1.43 | 47.4 | 13.9 |
| *Common*, medium loam | 0–20 | 26 | 2.63 | 50 | 1.22 | 53.6 | 21.6 |
| | 20–50 | 25 | 2.67 | 69 | 1.27 | 53.4 | 22.4 |
| | 50–100 | 38 | 2.69 | 89 | 1.35 | 49.8 | 19.2 |
| | 100–150 | 37 | 2.69 | 89 | 1.37 | 49.1 | 20.2 |
| | 150–200 | 18 | 2.70 | 22 | 1.37 | – | – |
| *Typical*, heavy loam | 0–20 | – | – | 8 | 1.21 | 59.7 | – |
| | 20–50 | – | – | 8 | 1.23 | 56.1 | – |
| | 50–100 | – | – | 10 | 1.35 | 53.2 | – |
| | 100–150 | – | – | 8 | 1.30 | 54.1 | – |
| *Leached*, heavy loam | 0–20 | 17 | 2.64 | 21 | 1.24 | 55.3 | 18.3 |
| | 20–50 | 19 | 2.65 | 31 | 1.32 | 52.8 | 17.7 |
| | 50–100 | 43 | 2.69 | 48 | 1.42 | 49.4 | 16.2 |
| | 100–150 | 46 | 2.71 | 44 | 1.42 | 48.0 | 15.0 |
| *Podzolized*, heavy loam | 0–20 | 4 | 2.63 | 4 | 1.24 | 55.7 | 26.9 |
| | 20–50 | 6 | 2.66 | 6 | 1.38 | 50.4 | 21.1 |
| | 50–100 | 10 | 2.71 | 10 | 1.52 | 45.4 | 14.0 |
| | 100–150 | 10 | 2.73 | 10 | 1.47 | 48.4 | 16.3 |
| *Xerophyte-wooded*, medium loam | 0–10 | 4 | 2.57 | 4 | 0.99 | 62.0 | – |
| | 20–30 | 4 | 2.59 | 4 | 1.02 | 61.0 | – |
| | 50–60 | 4 | 2.72 | 4 | 1.27 | 53.0 | – |
| | 90–100 | 4 | 2.70 | 4 | 1.26 | 54.0 | – |

The pore space plays many parts – all essential to the life of the soil: as the habitat of the soil biota, rooting space for plants, receptacle for water and nutrients in solution, and pathway for transport of dissolved and suspended materials in solution and suspension, and diffusion of gases (Berezin and Gudima 2002). The permeability of soils depends entirely on their porosity; the pore space is a fundamental attribute of self-organization – one of the soil's secrets that is still not well described or well understood but its importance for natural ecosystems and agriculture can be readily appreciated. Heavy traffic over the soil produces a compact layer (with a density of about 1.5 g/cm³ at 50 cm), which is not broken up by the plough and which impedes drainage and rooting (Ungureanu and Holmetchi 1990). A well-structured soil has some resistance to compaction but, even so, too much heavy machinery is being used on our soils.

Bulk density is only a crude indicator of porosity and structure but the technical description of porosity is hard to handle and of unproven utility. However, it is important to know that most of the water held in the soil and available to plants is held in fine pores *within* the soil aggregates. The coarser pores *between* aggregates allow for infiltration of water into the soil and drainage of any surplus. In general, the heavier the texture, the less the air-filled pore space but, in chernozem, it remains satisfactory for normal development and respiration of roots.

## Available Water Capacity

The maximum amount of water held in the soil and available to plants is taken to be the water content at cessation of drainage after the soil has been saturated, e.g. by prolonged heavy rainfall or irrigation. This is known as *field capacity*. The lower limit of plant-available water is known as *permanent wilting point*; water held more strongly than this, and not available to plants, is sometimes called *hygroscopic water*. Each of these parameters is used in calculating the soil water regime and is the object of attention and the subject of controversy in the standard works on soil water by Vysotsky, Lebedev, Stem and, more recently, Sheyin (1999).

In all chernozem, available water capacity is substantial, though rather less in the southern sub-types (Table 2). There are no data for *Xerophyte-wooded chernozem* but the values of the *Typical chernozem* may be applied in view of the small coefficient of variation. The topsoils hold 24–31% by volume of water at field capacity and 10–13% at wilting point. Subtraction gives an available water capacity of 16–19%, except for *Podzolized chernozem* (13.5%). These relatively high values are a result of both the soil texture (silt loam to clay loam) and the fine-granular structure – which takes us back to our earlier discussion of the hierarchy of organization in the soil (Fig. 6 of chapter "The Soil Cover"). The structure of the topsoil is also of crucial importance for the infiltration of rainfall and snow melt that might otherwise run off the surface and carry the topsoil with it.

The extraordinary aggregation and favourable pore-size distribution of pristine chernozem soils depend very much on their evolution under grassland and their stable humus content – which is gradually lost under the plough. An agronomic structural assessment according to the content of water-stable aggregates larger than 0.25 mm$^2$ reveals that plough layers have the lowest quality of structure; most commonly it is rated *unsatisfactory* (35–45% stable aggregates) or *bad* (strong dispersion) with less than 35% of stable aggregates; it is satisfactory, with 45–55% stable aggregates, only in layers deeper than 60–70 cm.

To compare the structure of pristine soil and ploughland, one of the authors made paired measurements on plots close to one another on the same landform (Table 3). In a *Xerophyte-wooded chernozem* and on ploughland cleared about 100 years ago, the highest quality structure extended to the depth of 70 cm, with 77–93% >0.25 mm aggregates and 48–81% >1 mm aggregates, which has been maintained, albeit somewhat degraded, during a 100 years under the plough. In a 20-year-old strip of forest on *Calcareous chernozem*, the grass cover did not compare with virgin steppe but, compared with ploughland, the amount of >0.25 mm aggregates in the topsoil is doubled and aggregates >1 mm increased 20-fold.

Dokuchaev called the Soroka chernozem "first class" not only for its fertility but also for its magnificent granular structure. That was in the past but we can be optimistic about the future because of the chernozem's self-repairing capacity; soil structure can be restored by good agricultural management.

**Table 2** Plant-available water in chernozem

| Chernozem sub-type and texture | Depth (cm) | Field capacity % volume | | Water content at permanent wilting point, % volume | |
|---|---|---|---|---|---|
| | | n | Mean | n | Mean |
| Calcareous, heavy loam | 0–20 | 23 | 28.2 | 29 | 12.3 |
| | 20–50 | 33 | 25.6 | 44 | 12.8 |
| | 50–100 | 53 | 24.2 | 72 | 11.3 |
| | 100–150 | 49 | 22.9 | 52 | 10.9 |
| Common, light loam | 0–20 | 6 | 31.0 | 8 | 12.7 |
| | 20–50 | 7 | 27.0 | 12 | 12.7 |
| | 50–100 | 15 | 23.9 | 20 | 13.2 |
| | 100–150 | 15 | 23.4 | 17 | 12.6 |
| Common, middle loam | 0–20 | 38 | 26.2 | 26 | 10.2 |
| | 20–50 | 44 | 24.4 | 46 | 10.9 |
| | 50–100 | 75 | 22.7 | 70 | 10.6 |
| | 100–150 | 65 | 21.1 | 62 | 8.6 |
| Typical, heavy loam | 0–20 | 4 | 30.2 | 4 | 12.1 |
| | 20–50 | 4 | 26.0 | 4 | 12.9 |
| | 50–100 | 5 | 24.5 | 5 | 12.5 |
| | 100–150 | 5 | 24.6 | 5 | 12.0 |
| Leached, heavy loam | 0–20 | 18 | 27.3 | 23 | 11.6 |
| | 20–50 | 24 | 25.5 | 29 | 12.2 |
| | 50–100 | 42 | 23.1 | 52 | 12.4 |
| | 100–150 | 37 | 22.7 | 42 | 11.5 |
| Podzolized, heavy loam | 0–20 | 4 | 24.6 | 4 | 11.1 |
| | 20–50 | 6 | 22.2 | 6 | 11.2 |
| | 50–100 | 10 | 21.2 | 10 | 12.1 |
| | 100–150 | 10 | 22.8 | 10 | 11.9 |

**Table 3** Comparison of structure of *Calcareous chernozem* and *Xerophyte-wooded chernozem* under forest and ploughland

| Paired profiles | Land use | Depth (cm) | Water-stable aggregates[a] (%) | |
| --- | --- | --- | --- | --- |
| | | | >0.25 mm | >1 mm |
| *Calcareous chernozem* | Arable | 0–15 | 20 | 1 |
| | Shelter belt | 0–15 | 38 | 20 |
| | Arable | 25–35 | 46 | 6 |
| | Shelter belt | 25–35 | 55 | 11 |
| | Arable | 45–55 | 45 | 10 |
| | Shelter belt | 45–55 | 44 | 8 |
| *Xerophyte-wooded chernozem* | Arable | 0–15 | 65 | 26 |
| | Forest | 0–15 | 93 | 81 |
| | Arable | 25–40 | 65 | 25 |
| | Forest | 25–40 | 82 | 56 |
| | Arable | 55–70 | 58 | 25 |
| | Forest | 55–70 | 77 | 48 |
| | Arable | 85–100 | 60 | 16 |
| | Forest | 85–100 | 62 | 17 |

[a]Determined by the method of Savinov, updated by the Russian Agrophysical Institute 1954.

Table 4, after Atamaniuk et al.[2] (1970), summarizes the physical properties of Moldavian chernozem that are most significant for farming. In every case, bulk density is satisfactory in the 20–100 cm layer – although close to the maximum desirable. Available water capacity is very high, ranging between 87 and 100 mm; even so, water scarcity is a constraint throughout most of the growing season in southern chernozem; water shortage is less pronounced in the north, which enjoys a higher rainfall (Lasse 1978).

## Drought

We shall not trespass on climatology but we need to mention drought. Lately, there has been a lot of talk about desertification – even in Moldova. We believe this to be an exaggeration although we may, certainly, speak about elements of desertification such as soil erosion and cutting down of shelter belts (Nour 2001; Ursu 2000). But drought is a regular visitor to Moldova and across tracts of chernozem everywhere.

Paleogeographer Mihailescu (1997) has assembled documentary evidence of 462 droughts in the western coastal steppes over the last 2000 years; drought recurs in cycles of 5–6, 11, 22, 33, 85–90, 180 and 300 years. Commonly, it is accompanied by black dust storms: in 1585, a chronicler recorded the drying-up of every spring, river and lake and "Because of the long absence of rains, there is so much dust that whole heaps of it are cast up under the fences, like snowdrifts ... Everywhere and everywhere, terrible famine has broken out" (Krupenikov 1974). Three centuries later, Grossul-Tolstoy (1856) and Orbinski (1884) in his sketch of productivity in the Province of Bessarabia wrote that droughts are uncommon in the north of the province but, in the south, folk expect 2 years of poor harvest and one very poor out of any 5 years owing to drought.

Closer to our own time and drawing upon state data, Constantinova (2001) has shown that drought occurs once in 10 years, in the north of the country; once in 5 or 6 years in the central part; and once in 3 years in the south. This distinct north–south zonation is also shown by our own data (Table 5): in zone 1, drought occurs 9.1 years in 29 (34%), including 3 years of severe drought (11%); in zone 2, in 10.1 years in 29 (35%) including 3.2 (15%) of severe drought; in zone 3, in 12 years in 29 (41%), including 5.5 (29%) of severe drought.

Drought is a fact of life on the steppe and such authorities as Zashiuk, Grosul-Tolstoy and Dokuchaev did not draw special attention to it; indeed, the outstanding Soviet geographer Berg wrote of Bessarabia as "the country with the fertile climate" (Krupenikov 1974). Daradur's 2001 monograph on rainfall variability over the last 120 years reports no evidence of progressive drying but he does note an increase in intense summer rainstorms – with more than 100 mm rainfall in a day.

Drought and soil erosion are two sides of the same coin. When soil permeability is impaired, intense rains bring destructive runoff that washes away the soil. The less the control of surface runoff, the more the erosion and the greater the likelihood of shortage of water for crops. Izmailisky, Kostichev, Vysotsky and others have established that repeated poor harvests are caused

---
[2]As described by Atamaniuk and others in *Soils of Moldova* vol. 1 (Ursu 1984).

**Table 4** Grouping of soils according to agrophysical properties

| Groups | Main soil types | Bulk density (g/cm³) by layer (cm) | | Mean stock of available water (mm) and air-filled porosity (%) by layer (cm) | | Notes |
|---|---|---|---|---|---|---|
| | | 20–50 | 20–100 | 20–50 | 20–100 | |
| I | *Common chernozem* and *Calcareous chernozem*, silt loam and heavy loam | <1.30 | <1.35 | 40 >18 | 100 >15 | Optimal compaction, insufficient soil water, very intensive gaseous exchange for the whole growing season |
| II | *Podzolized*, *Leached* and *Typical chernozem* | 1.30–1.35 | 1.35–1.40 | 37<br>17 | 87<br>15 | Compacted, average sufficiency of water |
| III | *Grey* and *Dark-grey forest* | 1.40–1.45 | | 38<br>15 | 86<br>12 | Compacted, especially unsatisfactory for the subsoil |

**Table 5** Moisture conditions of the growing season in Moldova

| Climatic zone | Meteorological station | Droughts 1960–1988[a] | | | | | | | | | |
|---|---|---|---|---|---|---|---|---|---|---|---|
| | | Very severe < 0.4 | | Severe 0.4–0.5 | | Moderate 0.5–0.6 | | Slight 0.6–0.7 | | Σ < 0.7 | |
| | | Years | % | Years | % | Years | % | Years | % | Years | % |
| I | Briceni | 3 | 10 | – | – | 1 | 3 | 2 | 7 | 6 | 21 |
| | Edinet | 4 | 14 | 2 | 7 | 2 | 7 | 2 | 7 | 10 | 35 |
| | Soroca | 5 | 17 | 1 | 3 | 2 | 7 | 1 | 3 | 9 | 31 |
| | Camenka | 5 | 17 | 1 | 3 | 1 | 3 | 2 | 7 | 9 | 31 |
| | Balti | 5 | 17 | 2 | 7 | 2 | 7 | 1 | 3 | 10 | 35 |
| | Ribnita | 6 | 21 | 2 | 7 | 1 | 3 | 2 | 7 | 11 | 38 |
| | Floresti | 5 | 17 | 2 | 7 | 1 | 3 | 1 | 3 | 9 | 31 |
| II | Bravicea | 5 | 17 | 1 | 3 | 2 | 7 | 2 | 7 | 10 | 35 |
| | Cornesti | 5 | 17 | 1 | 3 | 2 | 7 | 1 | 3 | 9 | 31 |
| | Carpineni | 5 | 17 | 2 | 7 | 2 | 7 | 1 | 3 | 10 | 35 |
| | Baltata | 4 | 14 | 1 | 3 | 2 | 7 | 4 | 14 | 11 | 38 |
| | Chisinau | 5 | 17 | 2 | 7 | 1 | 3 | 2 | 7 | 11 | 38 |
| III | Dubasari | 6 | 21 | 2 | 7 | 2 | 7 | 2 | 7 | 12 | 41 |
| | Tiraspol | 6 | 21 | 2 | 7 | 2 | 7 | 2 | 7 | 12 | 41 |
| | Stefan Voda | 7 | 24 | 2 | 7 | 1 | 3 | 2 | 7 | 12 | 41 |
| IV | Leovo | 5 | 17 | 3 | 10 | 2 | 7 | 3 | 10 | 12 | 41 |
| | Comrat | 7 | 24 | 1 | 3 | 2 | 7 | 2 | 7 | 13 | 45 |
| | Ceadir-Lunga | 6 | 21 | 2 | 7 | 1 | 3 | 2 | 7 | 11 | 38 |
| | Cahul | 7 | 24 | 1 | 3 | 2 | 7 | 1 | 3 | 11 | 38 |

[a]Drought defined according to Seleaninov's (1928) *hydrothermal quotient* (the ratio of precipitation to the sum of day-degrees above 10°C). The index ranges from <0.4 for very severe drought to 0.7–1.0 for near-drought conditions and >1.0 for adequate rainfall.

mainly by soil management and not by lack of rainfall (Kovali 1949). In other words, *agricultural drought* (lack of water in the soil) is more frequent than *meteorological drought* (a period of below average rainfall and too little to support the common field crops); and *political drought*, where various shortcomings are attributed to drought, is commonplace.

Drought is never welcome but those who come through best are those who care for their soil best (Kovali 1947, Timiriazev 1957). Drought can be mitigated by a prudent revival of supplementary irrigation but, also, by ecological agriculture; by selecting crop rotations that achieve the best possible ground cover and mulching bare soil – not only to protect the soil from rain splash but to arrest unproductive evaporation (Livovici 1986). There is no doubt that irrigation is an effective insurance against drought, but it is costly. We should first make the best use of the rain – by maintaining the soil as a free reservoir and promoting recharge of groundwater and springs by promoting deep percolation.

# References

Atamaniuk AK, RM Vladimir and LS Karapetean 1970 *Physical and ameliorative properties of soils in Moldova*. Stiinta, Chisinau, 70p (Russian)

Barasnina LN 1960 Accumulation and decomposition of soil organic matter from perennial leguminous crops, winter wheat and maize. 237–256 in MI Sidorov (editor) *Soil tillage in Moldavia*. Cartea Moldoveneasca, Kishinev (Russian)

Berezin PN and II Gudima 2002 Pore space in soils as living environment and basic factor for crop productivity. 210–212 in *Agriculture and ecology*. All-Russian Institute of Agriculture and Erosion Control, Kursk (Russian)

Bondarev AG 2002 Soil resilience and sensitivity to compaction under cultivation. 208–209 in *Soil resilience to natural and anthropogenic influence*. Moscow (Russian)

Constantinova TS 2001 Agrotechnical and phyto-ameliorative measures for minimisation of erosion. 142–185 in DD Nour (editor) *Soil erosion. The essence of the process, consequences, minimisation, stabilisation*. Pontos, Chisinau 427p (Russian)

Daradur M 2001 *Changes and evaluation of the risk of extreme conditions of humidity*. Institute of Geography, Academy of Sciences of Moldova, Chisinau, 160p (Russian)

Grossul-Tolstoy AI 1856 *Review of the rivers, soils and places in Novorossiyik region and Bessarabia regarding agriculture*, Odessa (Russian)

Jacks GV 1954 *Soil*. Nelson, Edinburgh, 221p

Kovali T 1949 Drought control. In: *The history of Russian agriculture*. Selihozgiz, 262p (Russian)

Krupenikov IA 1967 *Chernozem of Moldova*. Cartea Moldoveneasca, Kishinev, 427p (Russian)

Krupenikov IA 1974 *The influence of soils on yield and quality of tobacco (Review)*. Moldavian Research Institute of Technical-Economic Information, Kishinev, 45p (Russian)

Lasse GF 1978 in AB Lebedeva (editor) *Climate of SSR Moldavia*. Gidrometeoizdat, Leningrad, 374p (Russian)

Livovici MI 1986 *Water and life*. Mysl, Moscow, 253p (Russian)

Medvedev VV 1988 *Optimisation of agrophysical properties of chernozem soils*. Agropromizdat, Moscow, 160p (Russian)

Mihailescu C 1997 *Evolution of the geographic environment in the steppe zone of the NW part of the Black Sea and the rhythms of natural disasters during the last millennium*. Dr Hab. thesis, Institute of Geographical Sciences, Academy of Sciences of Moldova, Chisinau, abstract 44p (Romanian)

Nour DD (editor) 2001 *Soil erosion. The essence of the process, consequences, minimalisation, stabilisation*. Pontos, Chisinau, 427p (Russian)

Orbinski RV 1884 *Review of production forces in the Bessarabian Region*. SPb, Odessa, 166p (Russian)

Scerbacov AP and II Vassenev 2000 *Anthropogenic evolution of chernozem soils*. Voronej State University, Voronej, 411p (Russian)

Seleaninov GT 1928 On the evaluation of agroclimate. 165–177 in *Works on agricultural meteorology, Vol. 20*. Gidrometeoizdat, Leningrad (Russian)

Sinkevici ZA 1989 *Modern processes in chernozemic soils of Moldova*. Stiinta, Chisinau, 215p (Russian)

Sheyin EV 1999 On the particularities of agrophysical development of soils in Russia. *Pochvovedenie* 1, 49–53

Timiriazev KA 1957 *General agriculture and the physiology of crops*. State Publisher of Agricultural Literature, Moscow, 325p (Russian)

Ungureanu VG and AM Holmetchi 1990 *The problem of improvement and restoration of soil fertility for chernozem of Moldova*. Stiinta, Chisinau, 187 (Russian)

Ursu AT (editor) 1984 *Soils of Moldova. Vol.1 Genesis, ecology, classification and systematic description of soils*. Stiinta, Chisinau, 351p (Russian)

Ursu AT (editor) 2000 *Soil degradation and desertification*. Academy of Sciences of Moldova, Chisinau, 307p (Romanian)

Vladimir PM (editor) 1990 *Compacted soils in Moldova*. Stiinta, Chisinau (Russian)

# The Chernozem Family

**Abstract** Chernozem are well-drained soils with a thick, black topsoil characterized by an extraordinary accumulation of humus, granular structure that renders them friable even when dry and high base saturation, overlying a limey subsoil. They occupy some 230 million hectares in the steppes of Eurasia and Prairies of North America; they grade to chestnut soils (*Kastanozem*) in drier regions and to *Dark grey forest soils* (*Phaeozem*) in wetter areas. The subtypes of chernozem appear to be an evolutionary as well as geographic (south–north) sequence. Two evolutionary stems may be distinguished, illustrated in Moldova by the sequence in the south of the country from *Calcareous* to *Common chernozem* with a side branch to *Xerophyte-wooded chernozem*; and in the more humid north by the sequence from *Typical* to *Leached chernozem*, with a side branch to *Podzolized chernozem*. *Calcareous chernozem* (*Calcic chernozem* in the World Reference Base) is the least developed family member with carbonates, inherited from the loess parent material, throughout the soil profile; compared with other sub-types, it contains less humus and is less resilient to erosion by wind and water. The topsoil of *Common chernozem* (*Haplic chernozem*) has more humus, a stronger structure and is free of carbonates; in the subsoil there is secondary carbonate as pseudomycelia as well as calcite inherited from the parent material. Soils buried 2000 years ago beneath Trajan's Bank are both *Calcareous* and *Common chernozem*, just like today's soils, which suggests that the evolution from *Calcareous* to *Common chernozem* is slow. A variant under oak parkland (*Vermic chernozem*) exhibits a humus content as much as 12% in the topsoil (almost 500 t/ha in total) and extraordinary biological activity. *Typical chernozem* exhibit the characteristics of chernozem in their most pronounced form: very thick, black topsoil with intensive accumulation of mull humus and well-developed granular structure; carbonates are leached to below 50–90 cm and there is little secondary carbonate. Particle size distribution shows weak clay accumulation but the mineralogy indicates the beginnings of clay destruction. *Leached chernozem* are close to *Typical*, distinguished by leaching of carbonate to below 82/125 cm and the beginnings of silt and clay movement from the A to the B horizon. In *Podzolized chernozem*, carbonates are leached below 75/140 cm; the A horizon has about 85% base saturation, a pH in water of 5.8–6.1 but only 4.5–5.1 in dilute salt solution, reflecting 5–6 meq/100 g exchangeable acidity. All chernozem are very fertile. The bonitet rating compiled by statistical comparison of crop yields, soil type and texture takes the *Typical chernozem* as the gold standard of 100 points. Points are deducted according to attributes of soil and land that are associated with depression of crop yield compared with the standard; further points are deducted according to the degree of erosion.

**Keywords** Chernozem taxonomy – Calcareous · Common · Typical · Leached · Podzolized · Morphology · Micro-morphology · Bonitet rating

## Family Relationships

We have concluded the description of the chernozem as a whole – as a unique sub-system of the soil cover of Moldova and the world. In essence, chernozem are well-drained soils characterized by a thick, black topsoil that has an extraordinary accumulation of humus; a strong, fine-granular structure that renders the soil

**Fig. 1** *Calcareous chernozem* from the southern plain of Moldova: (**a**) (*right*) soil profile, virgin *Calcareous chernozem*; (**b**) (*left, above*) steppe with *Stipa* spp.; (**c**) (*left, below*) farmland (from Academician AT Ursu)

friable, even when dry, and an ion-exchange complex that is completely or almost completely saturated by bases, predominantly by calcium; overlying a lime-rich subsoil. In the *World Reference Base for Soil Resources* (IUSS WRB 2006), this translates as a *mollic horizon*[1] overlying, within 50 cm of the base of the mollic horizon, either a *calcic horizon*[2] or concentrations of secondary carbonate. In these early chapters, we have also drawn attention to differences in key attributes between the various members of the chernozem family. These family members are now described within the framework of the soil classification for Moldova by Krupenikov (1987); for international correlation, we also refer to the *World Reference Base for Soil Resources*.

Worldwide, chernozem occupy an estimated 230 million hectares – mainly in the mid-latitude steppes of Eurasia and Prairies of North America (Fig. 1). They share the distinctive, well-structured topsoil with *chestnut soils* or *Kastanozem*, into which they grade in the drier short-grass steppe of Eurasia and the Great Plains of North America. *Kastanozem* occupy twice the area of chernozem, some 465 million hectares; their topsoil is thinner and not so rich in humus, hence the chestnut rather than black colour, and the underlying lime-rich layer is more strongly developed. In more humid areas, chernozem grade into *Dark grey forest soils* (*Phaeozem* in the *World Reference Base*), which lack a lime-rich subsoil. *Phaeozem* occupy some 190 million hectares in the forest steppe in Europe, in northeast China, the Prairies of North America and Pampas of South America.

Many authors consider the sub-types of the chernozem to be an evolutionary as well as geographic sequence (Krupenikov 1967, 1992, Ursu 1985, Sinkevich 1989). The idea is supported by Alekseev's mineralogical work (1999) and by Chendev's recent investigation of soils buried under tumuli in southern Russia (Chendev 2001). Probably, two evolutionary stems may be distinguished: illustrated in Moldova by development from *Calcareous* to *Common chernozem*, with a side branch to *Xerophyte-wooded chernozem*, in the south of the country; and in the more humid north of the country, from *Typical* to *Leached chernozem*,

---

[1] A mollic horizon is at least 25 cm thick, with a structure strong enough to prevent the soil being hard and massive when dry, dark colour (Munsell chroma 3 or less when wet and 5 or less when dry), an organic carbon content of at least 0.6% and base saturation at least 50%.

[2] A calcic horizon is at least 15 cm thick, with a calcium carbonate equivalent content of at least 15% by volume and at least 5% by volume of secondary carbonates.

with a side branch to *Podzolized chernozem*. Members of the chernozem family and their characteristic vegetation are illustrated in Fig. 2 of chapter "The Soil Cover" and Figs. 1, 2, 3, 4, 5, 6 and 7.

## Calcareous Chernozem

(*Calcic chernozem* in WRB)

*Location*: Mainly in the south of the country; in the north and central parts, *Calcareous chernozem* are restricted to recent landforms such as river terraces.

*Identification*: Black earth topsoil (A horizon); carbonates throughout the soil profile giving an alkaline reaction throughout.

*Land suitability*: Because of drought, *Calcareous chernozem* has the lowest *bonitet rating* for arable crops (see section "Land Evaluation and Land Degradation"); sugar beet is not recommended because of the excess nitrate but the soil is rated 100 for vines.

Microscopic examination of thin sections (Fig. 8) reveals a *mull* type of humus, in which the organic matter is intimately associated with the clay fraction, and micro-aggregates of size 0.3–1.5 mm to a depth

**Fig. 2** Eroded *Ordinary chernozem*, southern plain of Moldova: (**a**) (*right*) soil profile; (**b**) (*left, above*) association of grasses; (**c**) (*left, below*) arable on sloping land (from Academician AT Ursu)

**Fig. 3** Typical *chernozem*, low-humic phase, on the southern plain of Moldova: (**a**) (*right*) soil profile; (**b**) (*left*) natural steppe vegetation with *Stipa* spp (from Academician AT Ursu)

**Fig. 4** *Typical chernozem*, Balti steppe, Moldova: (**a**) soil profile; (**b**) arable; (**c**) natural steppe vegetation (from Academician AT Ursu)

**Fig. 5** *Leached chernozem*, Codri, central Moldova: (**a**) (*right*) soil profile under cultivation and (**b**) (*left*) farmland (from Academician AT Ursu)

of 50–60 cm. Below this depth, the original fabric of the parent material prevails. Secondary carbonates occur as pore linings (*pseudomycelia*). The presence of felspar, mica and heavy minerals indicates weak weathering.

In all respects, *Calcareous chernozem* is the least developed member of the family. During the Quaternary Period, the landscape was coated by thick, limey loess (loams contain about 16% calcium carbonate, light clays about 12%). *Calcareous chernozem* still contain carbonate throughout the soil profile: about 2% at 0–20 cm; 7.5% at 50–60 cm; and 15% at 190–200 cm (Ionko 2002). Compared with other subtypes, they have less humus, with a closer humic:fulvic ratio, and lose this humus faster; they are less resilient to erosion by wind and water – so *Calcareous chernozem* are vulnerable soils and require careful management, particularly in respect of regular inputs of fresh organic matter.

A variant known as *Southern chernozem* occurs in SE Moldova, on the left bank of the River Dniester near Dniestrovsk and in southern Ukraine and the Crimean

**Fig. 6** *Leached chernozem* in the forest-steppe zone of northern; Moldova: (**a**) (*right*) Virgin *Leached chernozem* and (**b**) (*left*) *natural steppe vegetation* (from Academician AT Ursu)

steppe. These are loamy soils, rarely heavy loams, with less than 2% of humus in the topsoil. Their salient characteristic is the presence of gypsum below 2 m; it is prominent between 300 and 350 cm. Phosphate content is low throughout but they have favourable agrophysical properties – although in arable soils the structure is much deteriorated.

## Common Chernozem

(*Haplic chernozem* in WRB)

*Location*: South of Codri, this is the most widespread chernozem. It also occurs in the centre and north of the country on the highest river terraces and low interfluves.

*Identification*: Black earth topsoil (A horizon) free of carbonates and neutral in reaction. Effervescence with dilute hydrochloric acid is observed from a depth of 30–60 cm, where the reaction is alkaline. The accumulation of humus and expression of soil structure are stronger than in the *Calcareous chernozem*.

*Land evaluation*: The *bonitet* rating for field crops is 82 and for vineyards 100.

Microscopic examination of the topsoil (Fig. 9) shows, to a depth of 20 cm, well-defined microstructure with rounded aggregates 0.8–1.8 mm in size and a branching network of pores 0.2–0.4 mm diameter, with some up to 1.5 mm size. The mineral skeleton is mainly quartz and felspar, with little mica and heavy minerals. Organic matter includes pale brown plant residues and mull humus. The pattern is similar to a depth of 70–80 cm but with larger aggregates, amongst which worm casts are prominent. Secondary carbonate is precipitated as pore linings. Deeper in the soil profile, to about 2 m, the matrix is denser, the colour lighter, calcite pore linings are prominent and calcite crystals are visible throughout the soil matrix.

*Common chernozem* and *Calcareous chernozem* are similar in terms of aggregation and porosity but the *Common chernozem* has a more complex architecture, representing a more mature profile. Interestingly, the soils buried under Trajan's Bank some 2000 years ago are both *Calcareous* and *Common chernozem*, just like today's soils, which suggests that the evolution from *Calcareous chernozem* to *Common chernozem* is slow. However, I. Moshoia (1992) has demonstrated that this evolution may be accomplished by management – by intensive application of manure and straw, incorporating deep-rooting lucerne in the crop rotation and, above all, by irrigation.

## Xerophyte-Wooded Chernozem

(*Vermic chernozem* in WRB)

*Location*: In parkland dominated by pedunculate oak (*Quercus robur*), locally known as *ghimet*, with a rich ground flora of grasses and herbs. T. Zuikov has mapped 160 individual locations, a total of 19,000 ha, around the margins of the Tigecheskoy Heights and in Bender Wood, but not so far as the Dniester. This kind of vegetation is a geologically recent incomer, colonizing from the southwest along the Black Sea coast in the Holocene (Nicolaeva 1963). The adjacent soils are *Calcareous chernozem*

**Fig. 7** *Grey-wooded soil* in the forest-steppe zone of northern Moldova: (**a**) (*above, left*) soil profile; (**b**) (*above, right*) Oak woodland; (**c**) (*below*) farmland (from Academician AT Ursu)

(36%), *Ordinary chernozem* (18%) and *Leached chernozem* (15%).

*Identification*: The outstanding characteristic is the extraordinary humus content (Table 1) and conspicuous worm casts and crotovinas. The A horizon is about 45 cm thick; B1 45–69 cm; B2 69–96 cm. Effervescence with dilute hydrochloric acid increases from 30 cm down to 70 cm.

*Land suitability*: Under field crops, *bonitet* rating is 98.

In chapter "Life in the Soil", we mentioned the mass of roots in cultivated crops (between 4 and 6 t/ha of which 80% are concentrated in the 0–20 cm layer). Here the distribution is quite different. We may discount the roots of woody plants that contribute little to the accumulation of humus but in this ecosystem the roots of herbs constitute an amazing 17.5 t/ha in the topmost metre of the soil profile, as well as 10.4 t of dead residues. The humus itself amounts to 175 t/ha in the top 20 cm and 455 t/ha in the top metre – hardly

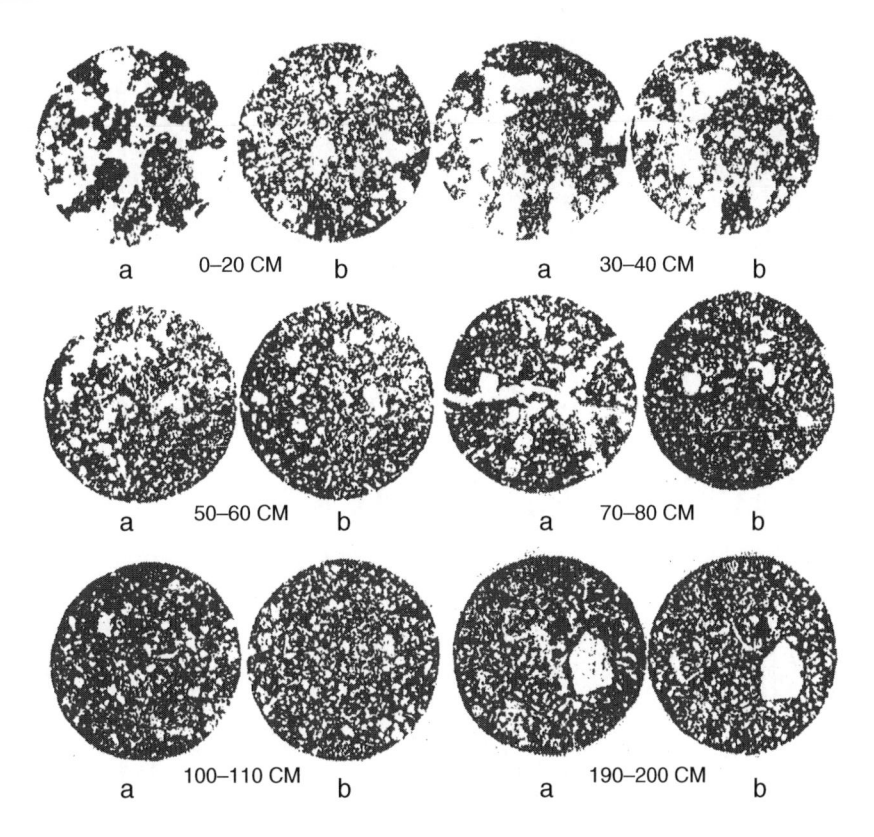

**Fig. 8** *Calcareous chernozem,* micromorphology: (**a**) direct light; (**b**) crossed nicols

in proportion to the mass of roots, which suggests very active turnover.

Moreover, every square metre has 242 invertebrates, more than 100 of them earthworms, compared with 38 and 14, respectively, in *Calcareous chernozem*; and 44 and 33, respectively, in *Ordinary chernozem*. Respiration rates are 12–16 volume% in *Xerophyte-wooded chernozem*, compared with 3.8–4.2% in *Common chernozem* and 0.8–1.4% in *Calcareous chernozem*. This remarkable biological activity is not maintained when the land is cleared for cultivation but the *Xerophyte-wooded chernozem* still exhibits a lot of resilience.

Unfortunately, we have no micromorphological examination for these soils but we may recall the robust soil structure described in Chapter 11: in the 0–25 cm layer, 93% of the soil mass is composed of water-stable aggregates >0.25 mm diameter (of which 81% are > 1 mm); in the 25–40 cm layer, 82% (45% > 1 mm) and in the 55–70 cm layer 77% (48% > 1 mm). This means rapid permeability and high available water capacity – so little loss of water by runoff or evaporation.

The geologically recent arrival of oak woodland and the geographical association of *Xerophyte-wooded chernozem* with *Calcareous chernozem* suggests evolution of this unique sub-type of chernozem under park woodland from *Calcareous chernozem* through increased root and invertebrate activity and leaching.

## Typical Chernozem

(*Haplic chernozem* in WRB)

*Location*: Widely distributed across the Balti Steppe, the left bank of the Dniester north of Ribnita and, in places, across the Codri Heights.

*Identification*: *Typical chernozem* are typical in the sense of expressing the most pronounced features of the chernozem: very thick, black earth topsoil exhibiting intensive accumulation of *mull* humus, friable

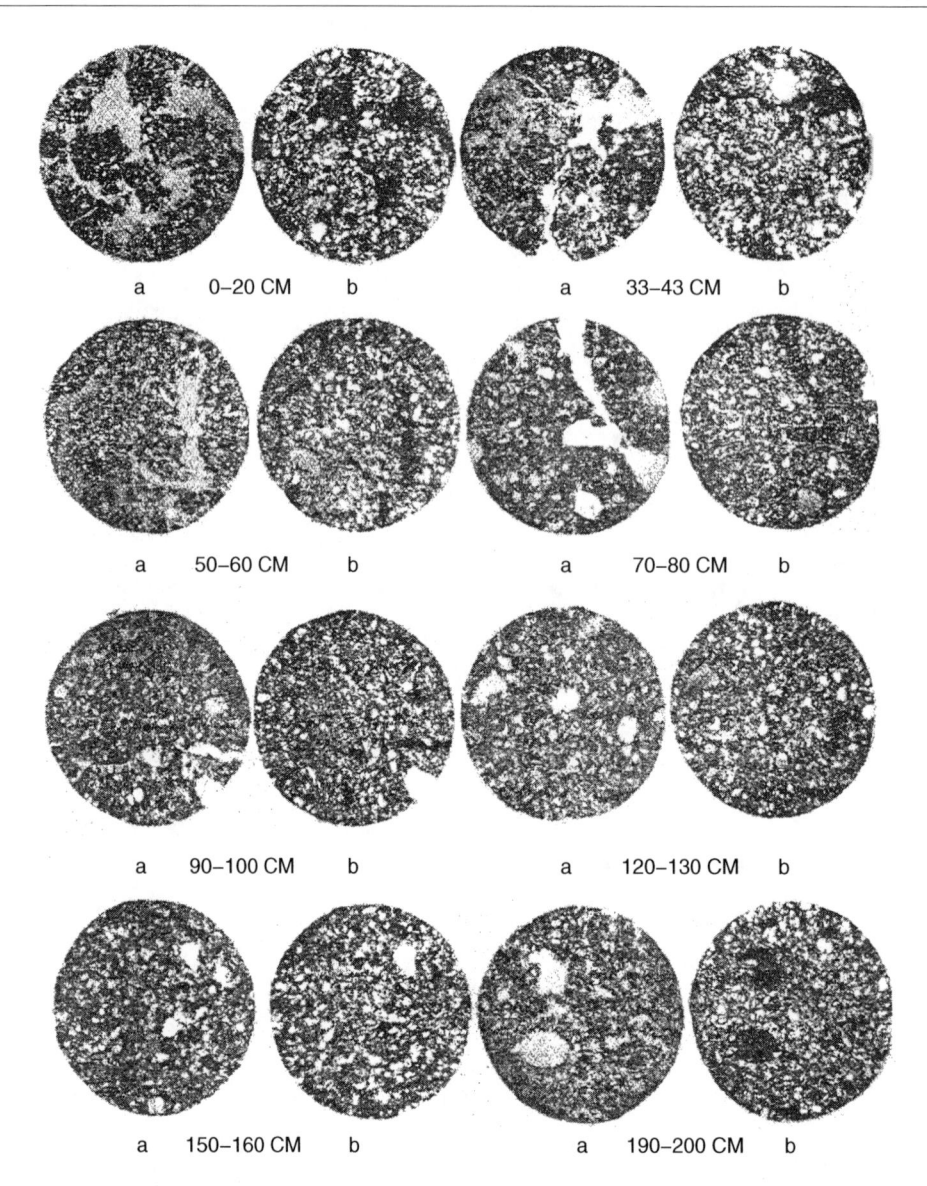

**Fig. 9** *Common chernozem*, micromorphology: (**a**) direct light; (**b**) crossed nicols

consistence and well-developed crumb or granular structure. Carbonates are leached to below 50–90 cm. The A horizon (36/47 cm) is weakly acid; the B1 horizon (36/47–54/70 cm) is close to neutral; and the B2 horizon (54/70–69/90 cm) shows some residual and secondary carbonates.

Particle size distribution shows weak clay accumulation in the soil profile while the mineralogy indicates the beginnings of clay destruction. The adsorption complex is 92–98% base saturated in the 0–50 cm layer (Krupenikov 1967), reflecting leaching of calcium and magnesium from the topsoil – which is a good thing since most field crops prefer a weakly acid reaction.

*Land suitability*: *Typical chernozem* is the absolute standard of 100 *bonitet* points. It is very suitable for all field crops, especially sugar beet, and fruits, especially stone fruits.

Compared with *Common chernozem*, the *Typical chernozem* has better developed micro-aggregates and pore architecture (Fig. 10). In the 0–60 cm layer, the mineral skeleton is mostly quartz and felspar; aggregates are rounded, defining a uniform network of pores; colour is dark brown in direct light and brownish black under crossed nicols. The 90–100 cm layer is lighter in colour, more closely packed and exhibits clumps of calcite crystals. Below 1 m,

**Table 1** *Xerophyte-wooded chernozem* roots and humus content, t/ha

| Soil layer (cm) | Living roots | | | Plant residues; humified and not humified | Soil humus stock |
|---|---|---|---|---|---|
| | Woody vegetation | Herbs | Total | | |
| 0–20 | 1.1 | 13.2 | 14.3 | 5.3 | 175 |
| 20–50 | 0.3 | 2.5 | 2.8 | 2.8 | 155 |
| 0–50 | 1.4 | 15.7 | 17.1 | 8.0 | 330 |
| 50–100 | 5.3 | 1.8 | 7.1 | 2.3 | 125 |
| 0–100 | 6.7 | 17.5 | 24.2 | 10.4 | 455 |
| 100–200 | 1.6 | 0.4 | 2.0 | No data | 40 |
| 0–200 | 8.3 | 17.9 | 26.2 | | 495 |
| 200–300 | 0.2 | 0.3 | 0.5 | | No data |
| 0–300 | 8.5 | 18.2 | 26.7 | | |

micro-aggregates are weakly developed. Throughout the profile, there is no oriented clay or clay skins and very little secondary carbonate; crystalline calcite is more common.

# Leached Chernozem

(*Haplic chernozem* in WRB)

*Location*: Widespread in the north of the country and across the heights of Soroca, Codri and Tigeci.

*Identification*: Geographically and in most attributes, *Leached chernozem* are close to *Typical chernozem*. They are distinguished mainly by deep leaching of carbonates – effervescence with hydrochloric acid occurs below 82–125 cm. The black topsoil (A horizon) is 36/46 cm thick and weakly acid, but still with base saturation of more than 90%; both the B1 (38/46–59/69 cm) and B2 (59/69–74/96 cm) show indications of movement of silt and clay down from the A horizon.

*Land suitability*: *Bonitet* rating 94 points.

Microscopic examination shows little difference from *Typical chernozem*, with well-developed micro-aggregation and porosity to great depth. Clumps of calcite are found at 115–125 and below (Fig. 11).

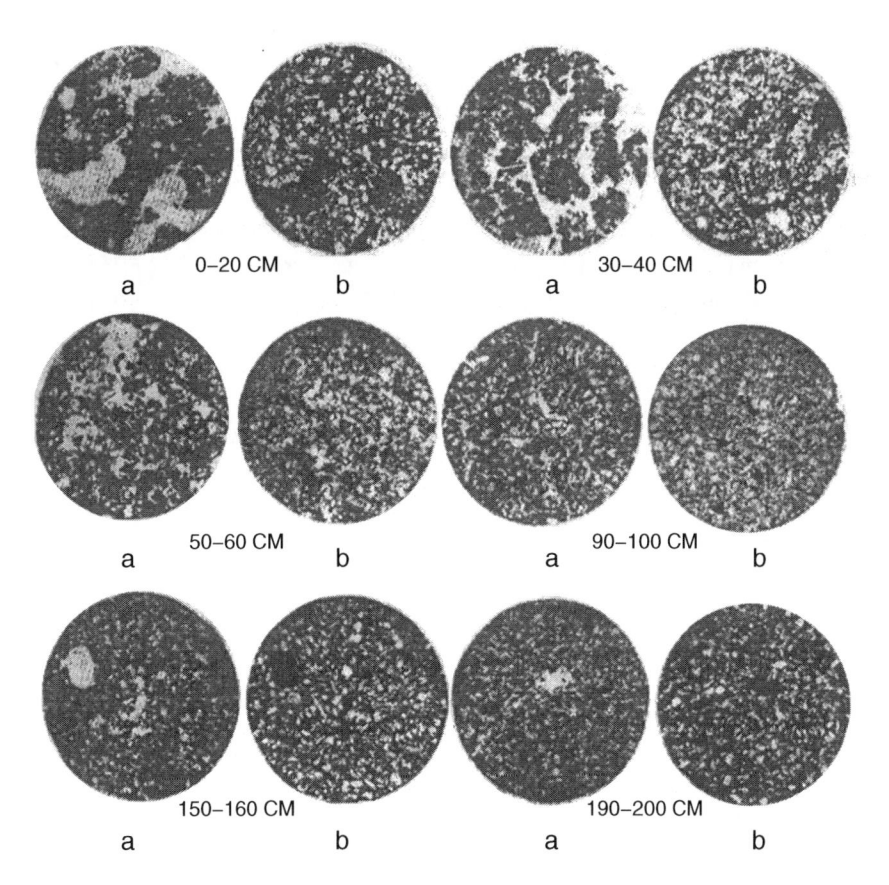

**Fig. 10** *Typical chernozem*, micromorphology: (**a**) direct light; (**b**) crossed nicols

**Fig. 11** *Leached chernozem*, micromorphology: (**a**) direct light; (**b**) crossed nicols

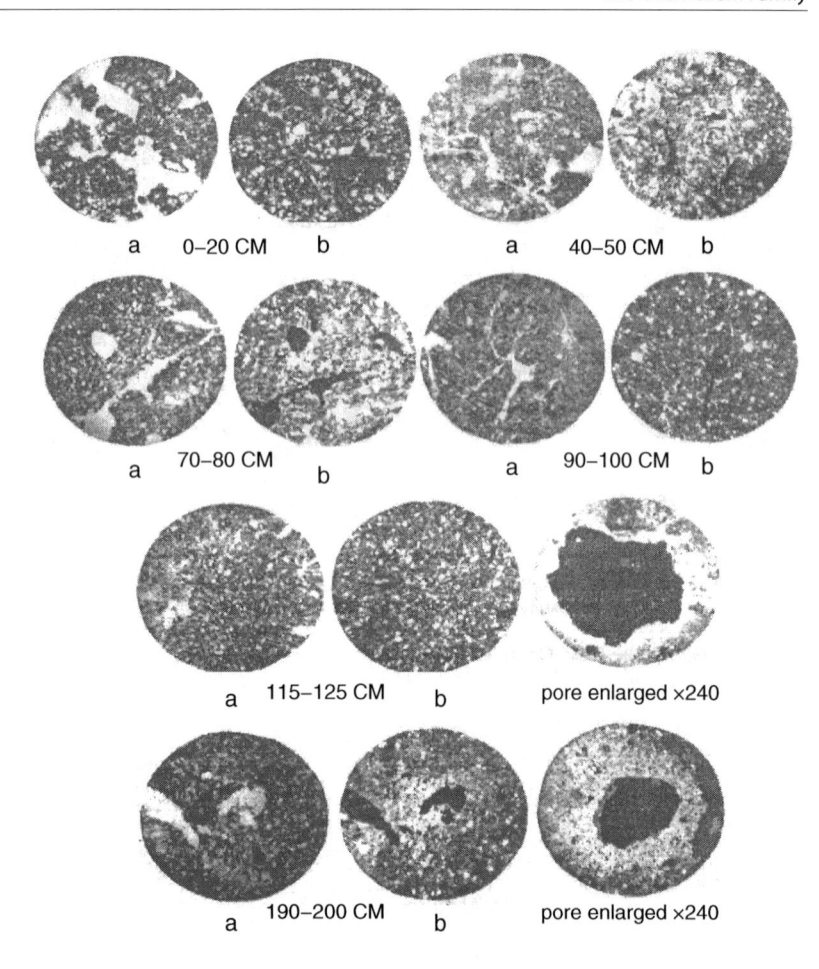

*Leached chernozem* may have developed for some time under forest that is now mostly receded or has been felled.

## Podzolized Chernozem

(*Luvic chernozem* in WRB)

*Location*: Northern Moldova, occurring close to the *Leached chernozem* but at higher elevations.

*Identification*: Carbonates are deeply leached; effervescence with dilute hydrochloric acid occurs below 75/140 cm. Compared with *Leached chernozem*, horizons are more clearly differentiated: A horizon to 32/43 cm, with 85% base saturation, a pH of 5.8–6.1 in water but 4.5–5.1 in dilute salt solution, reflecting 5–6 meq/100 g exchangeable acidity; B1 32/43–51/67; and B2 51/67–74/96.

*Land suitability*: *Podzolized chernozem* are very good soils for all field crops, potatoes and orchards. Sandy loam variants are sought after for cultivation of Virginia and Berley tobacco.

The regular altitudinal distribution of chernozem sub-types, first noticed by Dokuchaev (Krupenikov 1999) and subsequently confirmed by many others (Ursu 1985, 1999, Krupenikov 1967), is summarized in Table 2.

**Table 2** Altitudinal distribution of chernozem

| Chernozem sub-type | Absolute elevation, range (m) | Axis of elevation zone (m) |
|---|---|---|
| Calcareous | 30–200 | 115 |
| Common | 80–240 | 160 |
| Xerophyte forest | 140–200 | 190 |
| Typical | 180–220 | 200 |
| Leached | 180–240 | 210 |
| Podzolized | 225–245 | 235 |

**Table 3** *Bonitet* rating of the main soil types of Moldova

| Soils | Rating on soil properties | Correction coefficient | | | | | | | |
|---|---|---|---|---|---|---|---|---|---|
| | | For soil texture | | | | For erosion | | | For deep ploughing |
| | | Heavy loam | Middle loam | Light middle loam | Sandy | Low | Middle | High | |
| *Brown forest* | 72 | 0.9 | 0.8 | 0.7 | 0.7 | 0.8 | | | 0.9 |
| *Grey forest* | 68 | 0.9 | 0.8 | 0.7 | 0.7 | 0.8 | 0.7 | 0.5 | 0.8 |
| *Dark-grey forest* | 78 | 0.9 | 0.9 | 0.8 | 0.6 | 0.8 | 0.7 | 0.5 | 0.9 |
| Chernozem | | | | | | | | | |
|   Podzolized | 88 | 1.0 | 0.9 | | | 0.8 | 0.7 | 0.5 | |
|   Leached | 94 | 1.0 | 0.9 | 0.8 | 0.6 | 0.8 | 0.7 | 0.5 | |
|   Typical | 100 | 1.0 | 0.9 | | | 0.8 | 0.8 | 0.5 | |
|   Xerophyte-wooded | 98 | 1.0 | 0.9 | | | 0.8 | 0.6 | 0.5 | |
|   Common | 82 | 1.0 | 0.9 | 0.8 | 0.6 | 0.8 | 0.6 | 0.4 | 0.9 |
|   Calcareous | 71 | 1.0 | 0.9 | 0.8 | 0.6 | 0.8 | 0.6 | 0.4 | 0.8 |
|   Southern | 60 | 1.0 | 0.9 | 0.8 | | 0.8 | 0.6 | 0.4 | 0.7 |
| *Meadow* | 72 | 1.0 | 0.9 | | | | | | |
| *Meadow chernozem* | 100 | 1.0 | 0.9 | | | | | | |
| *River meadow* | 86 | 1.0 | | | | | | | |
| *Solonetz and solonchak* | 34 | 1.0 | | | | | | | |

## Land Evaluation and Land Degradation

The land evaluation or *bonitet* rating of soils has been compiled over 25 years of careful data acquisition and statistical comparison of crop yields, soil type and soil texture by L. Luneva, L. Ryabin and C. Markin. The gold standard of 100 points is awarded to *Typical chernozem* of fine loamy texture. Points are deducted according to attributes of soil and land that are associated with depression of crop yield compared with the standard, and further points are deducted according to the degree of erosion. Table 3 summarizes data for the main soils of the country as reported in the third volume of *Soils of Moldova* (Ursu 1986).

In every kind of chernozem, the fundamental attributes of the soil are ideal for farming and they can support continued, productive use under good management. But we cannot ignore the degradation and loss of humus under present farming systems and, in some cases, loss of the soil itself. This is a global issue but action to conserve and improve the soil has to be local – and within the framework of nation states. In our book *Erosion of Soils* (edited by Nour 2001) we listed various kinds of land degradation that affect Moldova and the whole of eastern Europe:

1. Loss of humus by biological processes
2. Un-compensated loss of nutrients
3. Chemical contamination
4. Soil compaction by heavy machinery and trampling by livestock
5. Irrigation-induced salinity and sodicity
6. Swamping of bottomlands
7. Loss of biodiversity
8. Erosion of topsoil by wind and water

Except in the case of soil erosion, the soil profile remains intact and a substantial recovery of soil functions is possible – at some expense – and this is the theme of the third part of this book. But soil eroded is lost forever.

## References

Alekseev VE 1999 *Mineralogy of soil formation in steppe and forest-steppe zones of Moldova: diagnosis, parameters, factors, processes.* Stiinta, Chisinau, 240p (Russian)

Chendev IG 2001 The problem of interrelation between the forest and the steppe in the light of archaeological data; the south Middle-Russian forest-steppe. 86–88 in IV Ivanov (editor) *The problems of soil evolution.* Biological Centre in Puschino, Academy of Sciences of Russia, Pushchino, (Russian)

Ionko OA 2002 Particularities of humus in carbonate chernozem from the Southern Russian Plain. 337–343 in AP Scerbacov (editor) *Modern problems of agriculture and ecology*. All Russian Institute of Agriculture and Erosion Control, Kursk, (Russian)

IUSS 2006 *World reference base for soil resources*. IUSS Working Group WRB, World Soil Resources Rept 103. UN Food and Agriculture Organisation, Rome, 128p

Krupenikov IA 1967 *Chernozem of Moldova*. Cartea Moldoveneasca, Chisinau, 427p (Russian)

Krupenikov IA 1987 Chernozem of Europe and Siberia: similarities and differences. *Pochovedenie* 11, 19–23

Krupenikov IA 1992 *Soil cover of Moldova*. Stiinta, Chisinau, 264p (Russian)

Nicolaeva LP 1963 *Forest from oaks in URSS Moldova*. Stiinta, Chisinev, 168p (Russian)

Nour DD (editor) 2001 *Soil erosion. The essence of the process. Consequences, minimalisation and stabilization*. Pontos, Chisinau, 427p (Russian)

Moshoia IG 1992 *The influence of irrigation on chernozem humus*. Thesis abstract Candidate in Agricultural Sciences, NN Sokolovski Research Institute of Soil Science and Agrochemistry, Kharkov 24p (Russian)

Sinkevici ZA 1989 *Modern processes in chernozemic soils of Moldova*. Stiinta, Chisinau, 215p (Russian)

Ursu AT (editor) 1985 *Soils of Moldova. Vol. 2 Soil geography, description of soil provinces, districts and sub-districts*. Stiinta, Chisinau 239p (Russian)

Ursu AT (editor) 1986 *Soils of Moldova. Vol. 3 Soil management, protection and improvement*. Stiinta, Chisinau, 333p (Russian)

# Accommodating Soil Diversity

**Abstract** Soils perform many functions including growing crops, water storage and regulation of stream flow and waste treatment. Fine distinctions within the chernozem family may be made according to many measurable attributes that affect the soil's performance in any of these functions. To match management practice with the soil, we need maps that show the distribution of these soil individuals. Examples of variability within single soil mapping units are given for *Calcareous* and *Typical chernozem* on two experimental stations. In-field soil variability is reflected in crop variability, e.g. in the Medvenskii District near Kursk, a threefold variation in the yield of winter wheat (1.8–6.0 t/ha) across a 4 ha plot; fourfold variation in the yield of barley (1.4–6.1 t/ha) across an 8 ha plot; and a twofold variation in sugar beet yield (11.0–24.2 t/ha) across a 63 ha field. Soil variation may be taken into account by precision farming, which makes use of farm machinery fitted with an on-board computer linked to the global positioning system and very detailed soil maps. This enables adjustment of seeding rate, application and mix of fertilizer across a field. Benefits include savings on fertilizer and avoidance of nutrient overload in soils and groundwater. Worldwide, there is great scope for precision farming and much may be achieved by intelligent adjustment of field boundaries and management practice to the soil pattern – even without sophisticated technology.

**Keywords** Soil variability · Detailed soil maps · Precision farming

## Soil Variability and Precision Farming

The soil performs many functions – such as accumulating and transforming energy from sunlight, water storage and regulation of stream flow, nutrient cycling and waste treatment. These functions depend on the interaction of various soil characteristics and it is hard to represent any complex attribute by a single numerical value, as we have seen with the complex derivation of the soil's *bonitet* rating.

We have now considered the chernozem as a special kind of soil and distinguished various sub-types according to particular features of their soil profile, such as the thickness of the topsoil and the depth to the calcareous layer. We also discussed how this diversity has arisen under the influence of a set of soil-forming factors: parent material, topography, climate and living organisms – each in itself varying in space and, except for the first, changing over time. Further distinctions may be made according to any number of measurable attributes; for instance, we may distinguish a class of *Common chernozem*, *heavy loamy*, *slightly eroded*. When it comes to land use planning and soil management – to match management practice with the soil – we need maps that show the distribution of these soil individuals.

Soil maps are made by examining many soil profiles and interpolating between observation sites by relating the attributes of each soil profile to its position in the landscape and, in particular, the factors of soil formation. We find that when any of these factors changes, the soil itself changes in phase with the changing factors. The soil surveyor predicts the characteristics of the soil at unvisited sites according to a model of the landscape built up from repeated observation, correlation with the soil-forming factors of the locality and testing with each successive observation.

Very good use can be made of conventional soil maps that represent the distribution of soil morphological classes – providing that the limitations of these maps are also understood. But to make very detailed soil maps of farms and individual fields, very intensive sampling is needed. The regularities in the landscape are underlain by local variability and the greater the precision we seek, either in the definition of the soil class, attribute or its numerical value, the more we find variability about the norm. For any measureable attribute, interpolation between measured sites may make use of spatial statistics (*geostatistics*) that were developed for gold prospecting. First, the rate of change of the attribute according to the distance between sampling points is established, then this relationship is used to predict the numerical value of that property for each point in the landscape. This is a prediction of probability – we cannot predict any numerical attribute exactly and, even in the case of direct measurement, there is always an element of random variation.

Alternatively, very detailed soil management maps may be made from observations of actual crop performance, that is, by their response to management, which can be done using large-scale air photos and yield-recording harvesters linked to the global positioning system.

The kinds of variability within what might be represented as a uniform soil mapping unit may be illustrated by some single-factor soil maps that we made in the 1960s with pedologist I. Shitihin. Figure 1 depicts Causheni State Testing Station, which is representative of *Calcareous chernozem*. The site is a plateau with slopes no greater than 1°, mapped as a single soil delineation at scale 1:2000. However, 20 auger borings to a depth of 150 cm revealed variation in the thickness of the humose layer between 80 and 100 cm; variation of the humus content in the plough layer between 3.6 and 4.9% about a mean 4.2%; variation in carbonate content between 0.9 and 15.6% about a mean of 2.5%, with available phosphorus inversely related to carbonate content; and variation of available potassium between 26 and 32 mg/100 g. These data underscore the need to take enough samples to establish a reliable numerical value for any soil characteristic, and they call into question the conventional procedure of determining attributes on a single, composite or bulked sample.

**Fig. 1** Detailed soil map of Causheni STS. Key: **I** (a) Location of experimental fields; o analysed soil profile pits, 111, etc.; **I–X** fields of rotation; (b) thickness of A+B horizons, cm: 1. 80–100, 2. 100–120; **II** Humus content, %: (**c**) plough layer, (**d**) subsoil: 1. <4, 2. 4–4.5, 3. >4.5; **III** CaCO$_3$, %: (**e**) plough layer, (**f**) subsoil: 1. <5, 2. 5–10, 3. >10; **IV** P$_2$O$_5$ content (Truog extract), mg/100 g: (**g**) plough layer, (**f**) subsoil: 1. 0.9–1.1, 2. 1.1–2.0, 3. 2.0–2.3; **V** K$_2$O (N ammonium acetate extract), mg/100 g: (**i**) plough layer, (**j**) subsoil: 1. 17–20, 2. 20–20, 3 >30

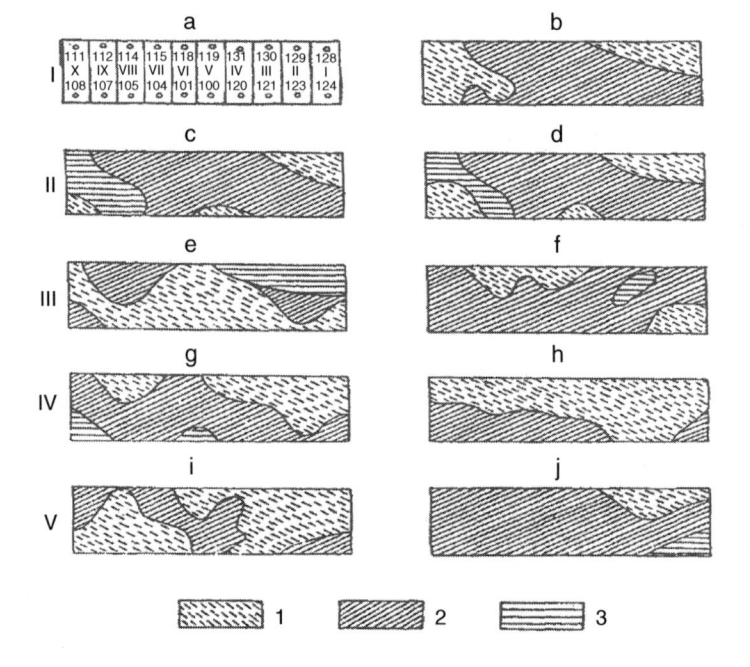

Figure 2 depicts Ribnita Soil Testing Station on *Typical chernozem*. Of the nine experimental fields, one-sixth of field 8 and one-fifth of field 9 are mapped as slightly eroded; all the rest is mapped as full-profile *Typical chernozem*. However, a similar investigation revealed variation in the thickness of the black topsoil between 100 and 120 cm about a mean of 110 cm, humus content in the plough layer of 4–5.3% about a mean of 4.7%, depth to carbonate between 60 and 90 cm about a mean of 75 cm, available phosphorus 8.4–19.8 mg/100 g about a mean of 13.5 mg/100 g and exchangeable potassium 31–56 mg/100 g about a mean of 36 mg/100 g.

Recently, I. Chendev and others (1999) have carried out a similar investigation of chernozem across the southern Russian Plain, with similar results for numerous sample points. We should expect that such significant soil variability would be reflected in crop variability – and it is. In-field variability has been measured by A. Scerbacov and I. Vassenev (2002) in the Medvenskii district near Kursk. They report a threefold variation in the yield of winter wheat (1.8–6.0 t/ha) across a 4 ha plot in 1998; fourfold variation in the yield of barley (1.4–6.1 t/ha) across an 8 ha plot in 2000; and a twofold variation in sugar beet yield (11.0–24.2 t/ha) across a 63 ha field in 1999. Even in a seed-replication plot with standard husbandry and enhanced fertility, the yields of wheat, barley and peas varied by a factor of 1.5. The authors explain crop variability in terms of variability in available water at

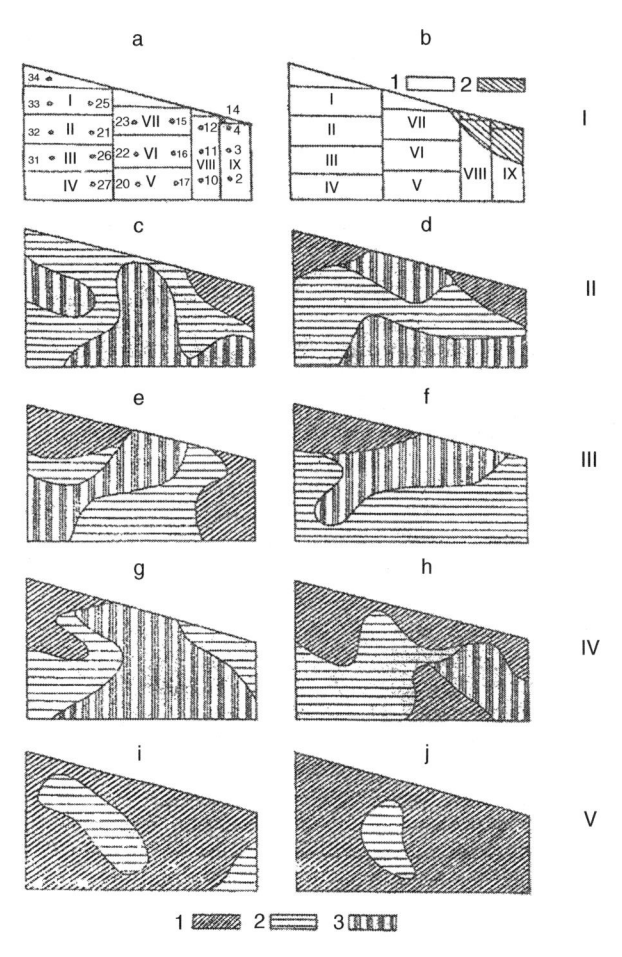

**Fig. 2** Detailed soil map of Ribnita STS. Key: **I** (**a**) layout of experimental fields; ○ site of measured soil profile pit; **I–IX** fields of rotation; (**b**) soil map: 1. *Typical chernozem*, 2. *Typical chernozem*, light clay, slightly eroded; **II** (**c**) thickness of A+B horizons, cm: 1. 90–100, 2. 100–120, 3. >120; (**d**) depth to effervescence with HCl, cm: 1. 55–65, 2. 65–80, 3. 80–90; **III** humus content, %, (**e**) topsoil, (**f**) subsoil: 1. 4.2–4.5, 2. 4.5–5.0, 3. >5; **IV** $P_2O_5$ (Truog extract), mg/100 g, (**g**) topsoil, (**h**) subsoil: 1. 8–10, 2. 10–14, 3. 14–20; **V** $K_2O$ (N ammonium acetate extract) mg/100 g; (**i**) topsoil, (**j**) subsoil: 1. 30–40, 2. >40

the time of flowering, available phosphorus and potassium and competition from weeds. They also suggest that soil variation should be taken into account in agronomic practice.

This is already common practice, under the name of precision farming, in the USA and western Europe. Farmers make use of machinery fitted with an onboard computer linked to the global positioning system and very detailed soil maps. This enables adjustment of seeding rate, application and mix of fertilizer across a field: see, amongst others, Walter (1996) and Parhomenko (1999) in the USA and Rogasik (2001) in Germany. The benefits include savings on fertilizer and avoidance of nutrient overload in soils and groundwater. Clearly, there is great scope for precision farming throughout eastern Europe. Even without sophisticated technology, much may be achieved by intelligent adjustment of field boundaries and cropping practices to the soil pattern.

# References

Chendev IG, PM Avramenco and TV Mascenco 1999 Microregional evaluation of variation in humus content of arable soils of the south-middle Russian forest-steppe. *Agrochemistry* 1, 24–28 (Russian)

Parkhomenkho EA 1999 Profitability analysis of precision agriculture. 37–43 in *Proceedings of 4th International Conference on Precision Agriculture*

Rogasik I 2001 *Grundlagen fur bodenvarialitat Kongress in Berlin* VDLUFA-Verlag (German), Berlin, 43p

Scerbacov AP and II Vassenev 2002 Tasks and perspectives for precision agriculturein Russia. 15–21 in *Modern problems of agriculture and ecology*. All Russian Institute of Agriculture and Erosion Control, Kursk (Russian)

Walter GD 1996 Site-specific sugar beet yield monitoring. 835–854 in *Proceedings of the 3rd International Conference on Precision Agriculture, Minneapolis*

# Society's Perspective

**Abstract** The soil is our life-support system. The natural system was self-contained, with a positive balance accruing to the soil and supporting its extraordinary biodiversity. Through trial and error, agriculture kept within the bounds of the self-contained ecological system for a long time in many places. In the past, many societies conserved their soils: recently soil has been taken for granted and the explosion of the human population and free-for-all capitalism have driven an intensification of farming that is degrading the land and soil across a quarter of the land surface; this degradation is cumulative. An immediate task of ecological agriculture is to arrest the processes of land degradation, turn them around and rebuild the soil. It will be easiest and most rewarding to begin with full-profiled chernozem. For eroded soils, the first step is to stop the erosion but raising their fertility to the level of the full-profiled soils will take a long time. Many of the technical issues are well understood. The societal issues are (1) better comprehension of the scientific and technical issues by the farmers themselves; (2) better interaction between soil scientists and others beyond the agricultural field; and (3) wider propagation and appreciation of the value of soils, and soil science, in society.

**Keywords** Soils and society · Land degradation · Ecological agriculture

## Soils and Society

We have described the chernozem as a unique kind of soil; in many ways it behaves like other kinds of soil but in some ways it is much more active. From its parent material, the blanket of calcareous loess that overlies the country rock, it inherits its texture, mineralogy and the stable chemical attributes that provide such a fertile framework for life – not least the plentiful supply of plant nutrients. As the medium of interaction with the atmosphere, it also receives a flux of carbon and nitrogen and builds up a rich store of these life elements. None of these life processes can take place without water. Here again, the soil is a working interface. Without the soil, exchange of water between the atmosphere and oceans would be very simple and very fast; some of the rain falling on the barren land would evaporate but most would run off directly to the ocean. Instead, the soil regulates this flow. It holds more than 60% of rainfall – to be used by plants, allowing less than 30% to run off directly to streams, while the balance recharges groundwater and stream base flow (Falkenmark and Rockström 2004).

In all these ways, the soil is our life-support system. The natural system was self-contained, with a positive balance accruing to the soil and supporting its extraordinary biodiversity. Some 12–15 thousand years ago, it assumed a more direct role with the birth of agriculture – and the productivity of human labour immediately increased 10-fold. In due course, farming became more complex; manual labour was supplemented by animal traction and later by machinery; rainfall was supplemented by irrigation, and manure was supplemented or replaced by artificial fertilizers. But the far-reaching application of technology has been far from benign.

In antiquity, farmers likened the soil to a woman and believed that it was tended by some special goddess – Isida of the Egyptians, Aster of the Phoenicians, Inana in Babylon, Artemis and Aphrodite in Ancient Greece, Venus in Ancient Rome. In his *Dialogues*, Plato replies to the question whether soil is like a woman or a woman like the soil: "Certainly, the woman is like the soil". The Romans conceived a variety of specialist gods – even a god of dunging, Scircus. And soil itself was honoured; the Sumarians believed that the gods made the first people from warm, moist soil; they could not have found a better material. But whereas polytheistic religions have been soil-friendly, there is little or nothing about soil conservation in the Bible or Koran.

Through trial and error, agriculture kept within the bounds of the self-contained ecological system for a long time in many places. Strict rules of cultivation were laid down in China and Japan; in Europe a system of diligent manuring developed in the forest zone and a quite stable farming system developed on the steppes; and in many tropical areas fertility was maintained by low-intensity shifting cultivation. The recent explosion of the human population and the rise of a free-for-all capitalism have driven intensification of farming which, for all its undoubted successes, has not always been accompanied by the necessary care of the soil. In the march of progress, the soil has been taken for granted.

Nowadays, considering the agricultural crisis throughout eastern Europe, comparisons are often made with western Europe and North America. Certainly, much can be learned from the West but all is not well there either. There are many examples of subsidized surplus production and not a few of land degradation, contaminated soil and polluted water. Indeed, land degradation is a global issue. Over the last quarter century and allowing for climatic variability, production has been declining across a quarter of the land surface and more than 20% of cropland (Bai and others 2008). And this degradation is cumulative. The previous global assessment, presented in 1990, estimated that 15% of the land was degraded (Oldeman and others 1990): and the degraded areas mapped then hardly overlap with those affected in the last 25 years. We should be very thankful for the resilience of our chernozem!

## Agricultural Priorities

The soil part of this book is concluded. Each chapter provides a basis for some decision that has to be made in establishing sustainable farming systems. Those attributes of the chernozem that remain constant, such as soil texture and mineralogy, provide the foundation for farming but require no improvement. In contrast, attributes that depend on the ecology of the soil, which have been degraded in recent decades and continue to degrade, present an urgent challenge. These ecological attributes are the humus and nutrient status, the biota, soil structure and all the functions that depend on them – including permeability, available water capacity and resistance to erosion.

An immediate task of ecological agriculture is to arrest the processes of land degradation, turn them around and rebuild the soil. It will be easiest and most rewarding to begin with full-profiled chernozem. For eroded soils, the first step is to stop the erosion. But raising their fertility to the level of the full-profiled soils will take a long time.

Land reclamation involves infilling of gullies, grading the slopes and initiating proper ecological practice on the newly mixed soil parent material; this may be undertaken as a local imperative and technical specifications have been developed (Krupenikov 2001, Krupenikov and others 2002) but it is not the first national priority. The same applies to reclamation of valley bottomlands that have become saline, sodic or boggy; amelioration is laborious and success uncertain. The national priority – statutory, administrative, financial and agronomic – must be conservation and improvement of full-profiled chernozem.

We have already mentioned the mighty effort in soil conservation made by the USA following the tragedy of the Dust Bowl in the 1920s and 1930s. Yet, even now, the nationwide Soil Conservation Service (recently re-named the Natural Resources Conservation Service) complains about a lack of understanding by both the authorities and the general public (Sposito and Reginato 1992). It lists its main tasks at the turn of the twenty-first century as follows:

1. Better comprehension of fundamental biophysical laws by those actually working with the land
2. Better interaction between soil scientists and others beyond the agricultural field
3. Wider propagation and appreciation of the value of soils, and soil science, in society

Truly, we have to say, with Einstein "As much as we know, as little we understand".

# References

Bai ZG, DL Dent, MA Schaepman and L Olsson 2008 Proxy assessment of land degradation. *Soil Use and Management* 24, 223–234

Falkenmark M and J Rockström 2004 *Balancing water and nature*. Earthscan, London

Krupenikov IA 2001 *Recommendation for improvement of eroded soils by bringing in soil materials*. Research Institute of Soil Science and Agrochemistry, Chisinau, 16p (Romanian)

Krupenikov IA, IS Constantinov and GP Dobrovolski 2002 *Degraded soils and their restoration by using natural ameliorants: review information*. Institute of Economics and Information, Chisinau 50p (Russian)

Oldeman LR, HTA Hakkeling and WG Sombroek 1990 *World map of the status of human-induced soil degradation; an explanatory note*. ISRIC, Wageningen and UNEP, Nairobi, 27p and 3 maps

Sposito G and RJ Reginato (editors) 1992 *Opportunities in basic soil science research*. Soil Science Society of America, Madison, WI

# Part III
# Principles of Ecological Agriculture

# Biological Cycles

**Abstract** Soil fertility depends on processes of both decay and synthesis. In natural ecosystems, a large and diverse population of living organisms well attuned to their environment accomplishes the conservative cycling of energy and nutrients. Turnover of the more easily decomposed fractions of soil organic matter takes only a few months to a few years, but turnover of recalcitrant fractions occurs over hundreds or thousands of years. Farming systems have reversed the natural accumulation of carbon in soils; the energy gap, created by the export of large amounts of biomass off-farm, is inexorably depleting the reserves of soil organic matter that have been built up over millennia. In natural ecosystems, pathogens are held in check by competition between many species in the food web. Agroecosystems are much simpler, less diverse, usually less well adapted to the environment and prey to outbreaks of weeds, pests and disease. Modern farming systems have replaced dependence on biological cycles by application of artificial fertilizer, chemical herbicides and pesticides. However, mineralization of organic matter, especially the more-easily decomposed or labile organic matter, still contributes a big share of crop nutrients. Nitrogen, in particular, is often released in excess of crop requirements and contaminates the groundwater, streams and lakes.

**Keywords** Soil fertility · Ecosystems and agroecosystems · Energy and nutrient budgets

## Biological Cycling of Soil Organic Matter

The organic part of soils, embracing living organisms and decomposing organic residues including humus, makes up just 3–6% of the mass of arable soils but it determines their activity and fertility. Living organisms comprise only 5–15% of the organic component but produce all the rest. The food chain includes amongst many: bacteria that feed on root exudates and which are, in turn, prey to protozoa and nematodes; fungi that decompose more complex organic materials; and arthropods that contribute to decomposition by comminuting plant residues, and earthwarms that mix and move organic matter and mineral soil.

The breakdown and synthesis of organic matter is accomplished through the simultaneous interaction of all the biota; the richer the biota, the more fertile the soil. Breakdown of the more easily decomposed fractions releases soluble plant nutrients that are rapidly taken up by both plant roots and microorganisms; losses of soluble nutrients through leaching are restricted by their brief residence time in the soil. At the same time, new organic compounds are synthesized, notably gums that stabilize soil structure.

The rate of decay of the various constituents of organic matter is by no means uniform. Parton and his colleagues (Parton et al. 1996), working in Illinois, estimated the residence times of easily decomposed, intermediate and resistant fractions of soil organic matter as 1–5, 20–50 and about 3000 years, respectively. Earlier, Jenkinson and Rayner (1977) at Rothamsted estimated the mean residence times of these same fractions as 2, 5 and 1980 years.

If soil biodiversity is the guardian of soil fertility and the health of the soil and crops, then frequent additions of fresh organic matter are the guardians of soil biodiversity. The need to fuel biodiversity in farming systems may be appreciated through a comparison of managed agroecosystems with natural ecosystems

**Table 1** Distinctions between natural ecosystems and agroecosystems

| Indices | Natural ecosystem | Agroecosystem |
|---|---|---|
| Productivity | Average | High to low |
| Trophic interaction | Complex | Simple, linear |
| Species diversity | High | Low |
| Genetic diversity | High | Low |
| Turnover of nutrients | Closed | Open |
| Stability | High | Low |
| Human dependence | Independent | Dependent |
| Sustainability over time | Long | Short |

(Table 1, after Gliessman 2000). Natural ecosystems are more complex and, therefore, more resilient but the resilience of agroecosystems can be augmented by maintaining and feeding the soil biota.

## Energy Balance of Arable Systems

Our own research using long-term field experiments with crop rotations and continuous cropping systems testifies to the energy gap in agroecosystems, especially under continuous cropping and with a high proportion of row crops in the rotation (Boincean 1999, Table 2). Although the fossil fuel energy expended in farming systems is much less than the solar energy accumulated by the crops, the greater part of the biomass energy produced by the crop is carried away from the field and lost to the system. Amongst the big issues for sustainable agriculture are to reduce the need for non-renewable energy and, at the same time, find ways of cheaply returning biomass energy to the soil.

In fact, soil organic matter makes up about three-quarters of global stocks of organic carbon (Parton and others 1988). This value is rarely computed and, as we have discussed in Chapter 8, probably underestimated but it is clear that soil organic matter plays a critical regulating role in the Earth's carbon cycle. Soil organic matter is the biggest carbon store that we know how to manage and the cumulative emissions of carbon dioxide from loss of soil organic matter are of the same order as emissions from burning fossil fuels.

The shift from mainly perennial vegetation in natural ecosystems to mainly annual crops also changes the distribution of biomass and soil organic matter. Chesniak's data for pristine Russian chernozem show a net primary productivity of 15.4 t/ha, of which the below-ground fraction constituted nearly three-quarters (11.2 t/ha). By contrast, under a rotation of arable crops (sugar beet, cereal and maize silage) the above-ground biomass amounted to 11.3 t/ha and below-ground 3.6 t/ha, less than one-third of the total (Chesniak and others 1983). In terms of the mass of roots, virgin soil surpasses ploughland more than threefold which, again, reminds us of the need for continual additions of organic matter to arable soils.

## Nutrient Balance of Arable Systems

We also need to consider the C:N ratio of the organic matter. Adding a lot of nitrogen-rich material encourages rapid decomposition whereas adding material

**Table 2** Energy balance of long-term field experiments at Selectia Experimental Station, Balti, 1962–1991

| Variants | | | Output (mJ/ha) | | | Input (mJ/ha) | | | Balance | |
|---|---|---|---|---|---|---|---|---|---|---|
| | | | Taken up by above-ground biomass | Non-compensated deficit of soil organic matter | Total | With crop residue | With manure | Total | mJ/ha | Annual energy deficit (%) |
| Rotations, % row crops | | 40 | 99.3 | 10.4 | 109.7 | 64.8 | 1.9 | 66.7 | −43.0 | 39.2 |
| | | 50 | 95.2 | 12.4 | 107.6 | 50.4 | 0.3 | 50.7 | −56.9 | 52.9 |
| | | 60 | 116.9 | 11.5 | 128.4 | 60.2 | 3.7 | 63.9 | −64.5 | 50.2 |
| | | 70 | 112.6 | 11.5 | 124.1 | 63.6 | 2.4 | 66.0 | −58.1 | 46.8 |
| Continuous crops | Corn for grain | Unfertilized | 100.4 | 20.0 | 120.4 | 35.8 | – | 35.8 | −84.6 | 70.3 |
| | | Fertilized | 141.0 | 16.8 | 157.8 | 51.1 | 4.4 | 55.5 | −102.3 | 64.8 |
| | Winter wheat | Unfertilized | 57.4 | 20.0 | 77.4 | 21.0 | – | 21.0 | −56.4 | 72.9 |
| | | Fertilized | 80.3 | 15.2 | 95.5 | 29.4 | 4.4 | 33.8 | −61.7 | 64.6 |
| Bare fallow | | Unfertilized | | 32.7 | 32.7 | 0 | 0 | 0 | −32.7 | 100 |
| | | Fertilized | | 26.9 | 26.9 | 0 | 4.4 | 4.4 | −22.5 | 83.6 |

with a wide C:N ratio results in fixation of nitrogen in the soil, so that it is unavailable to crops – at least in the season of manuring. All experience points to the importance of making good farmyard manure, which promotes conservation of soil organic matter – both as a source of plant nutrients and to mitigate the emission of greenhouse gases to the atmosphere (Rees and others 2001). Unfortunately, this truth was overlooked during the era of cheap artificial fertilizers.

The gross stock of organic matter is not always a reliable indicator of its agronomic value; in general, fresh organic matter is more active than highly decomposed material. Differential thermal and differential thermo-gravimetric analyses of humic acids separated from soils under the long-term field experiments at Balti indicate that, under arable crops with insufficient manuring, there has been a decrease in the most active, agronomically valuable fraction of soil organic matter – the so-called *labile* organic matter (Table 3).

Comparing crop rotations comprising 50% field crops and a low application of farmyard manure with rotations comprising 60% field crops but with a higher application of manure, we may note that crops took up 7.4% of the total labile fraction in the first case, compared with 14.4% in the second case. The conventional assumption is that the content of nitrogen in the labile fractions of soil organic matter is a good indicator of crop-available nitrogen; these results suggest that this assumption may be unjustified.

The use of nitrogen from organic matter is an important consideration for ecological agriculture. If nitrogenous fertilizer is not used, then the soil organic matter becomes the main source of nitrogen for crops. Cited data on decreasing productivity in agricultural systems are frequently linked to insufficient soil capacity to supply plants with nitrogen – so the transition to ecological systems of agriculture must begin with improvement of this capacity. Data from the Institute for Soil Science and Agrochemistry under S.V. Andries confirm the attrition of soil nitrogen stocks in Moldova: in 1989, a low content of nitrate nitrogen (<60 kg $NO_3$/ha in the upper metre) was recorded for 9% of wheat fields in Moldova; in 2000–2002, 69–93% of wheat fields were so characterized (Andreis and others 2001). Doubtless, the efficiency of fertilizers on such crops is very high; each kilogram of nitrogen fertilizer at optimal dose (60 kg N/ha) returns 16–25 kg of grain.

Data from the long-term field experiments at Selectia Experimental Station (Table 4) indicate that mineralization of fresh plant residues is the source of most available nitrogen and regulates the balance of soil organic matter. This means that fresh organic matter must be continually replenished and its rate of decomposition must be controlled. If we are to maintain the productivity of arable land and avoid both contamination of groundwater and excessive emissions of greenhouse gases, we need a simple way to assess and monitor soil organic carbon and nitrogen so that we may

- determine the maximum yield of crops in a rotation;
- optimize application of manure and fertilizers according to the contribution of organic and artificial nitrogen available to the crop;
- increase the efficiency of artificial fertilizers; and
- avoid losses of both organic and artificial nitrogen through leaching.

The model allows determination of the total inputs of nitrogen from different sources (crop residues, manure, artificial fertilizer, biological fixation, etc.) taking into account the amount available for crops. The model is validated by real data from field experiments

**Table 3** Mineralization of labile soil organic matter under crop rotations at Selectia Experimental Station, Balti, 1993

| Proportion of row crops in rotation (%) | Stock of labile organic matter (t/ha[a]) | | Yield of winter rye (green mass) (t/ha) | Nitrogen taken up by rye (kg/ha) | Nitrogen mineralized from soil organic matter (kg/ha[b]) | Mineralized labile fraction of soil organic matter (%) |
|---|---|---|---|---|---|---|
| | Carbon | Nitrogen | | | | |
| 40 | 19.5 | 0.50 | 5.4 | 49 | 98 | 19.6 |
| 50 | 16.4 | 0.86 | 3.8 | 32 | 64 | 7.4 |
| 60 | 26.8 | 0.79 | 6.2 | 57 | 114 | 14.4 |
| 70 | 15.5 | 1.02 | 4.3 | 35 | 70 | 6.9 |

[a]Labile soil organic matter determined according to Cambardella and Elliot (1992).
[b]Assuming 50% coefficient for use of nitrogen from soil organic matter.

**Table 4** Mineralization of soil organic matter (SOM) at Selectia Experimental Station, 1982–1991, mean values

| Experimental treatments | | Annual inputs of crop residues and manure | | Annual SOM deficit (t/ha/year) | Annual expenditure of SOM (t/ha/year) | Coefficient of mineralization of SOM (%) | % fresh residues in mineralization losses (%) |
|---|---|---|---|---|---|---|---|
| | | t/ha/year | Mineralization in first year (t/ha) | | | | |
| Bare fallow | a | 1.7 | 0.8 | 0.68 | 1.48 | 2.5 | 0.0 |
| | b | – | – | 0.82 | 0.82 | 1.4 | 54.0 |
| Continuous wheat | a | 3.3 | 2.0 | 0.38 | 2.38 | 3.5 | 84.0 |
| | b | 0.9 | 0.6 | 0.50 | 1.10 | 1.8 | 54.5 |
| Continuous maize | a | 4.0 | 2.5 | 0.42 | 2.92 | 4.3 | 85.6 |
| | b | 1.3 | 0.9 | 0.50 | 1.40 | 2.2 | 64.3 |
| Crop rotations, % row crops | 40 | 3.1 | 2.0 | 0.26 | 2.26 | 3.3 | 88.5 |
| | 50 | 2.2 | 1.5 | 0.31 | 1.81 | 2.6 | 82.9 |
| | 60 | 3.9 | 2.4 | 0.29 | 2.69 | 3.8 | 89.2 |

a, Fertilized; b, unfertilized.

at Selectia Experimental Station at Balti (Boincean 1999).

We find nitrates leaching to the groundwater under arable crops, even on unfertilized soils, so there is need to improve the cropping system at the design stage (Boincean et al. 2010). Systematic application of this procedure will need the active collaboration of farmers and agronomists and extension of the network of experimental sites at which the relevant parameters are measured – in particular, the differences in crops' uptake of nitrogen between fertilized and unfertilized sites. Information on the capacity of the soil to provide each crop and crop rotation with nitrogen is especially important in changing to ecological systems of agriculture. But each and every farmer will benefit, no matter what system is practised, because it is directed to improvement of soil health and the optimal use of increasingly expensive fertilizer.

Annual changes in soil organic carbon and nitrogen are immeasurably small compared with the large total stock. However, at the research level they can be followed using radionuclide tracers $^{14}$C and $^{15}$N which, also, enable us to track the transformation of organic matter added to the soil. The calculation is simple; the results are confirmed by long-term field experiments and farmers' experience and provide a tangible indicator of the fitness for purpose of farming systems and necessary corrections.

In ways like this we may prevent, rather than struggle to cure, the harmful consequences of interference with nature. In conventional farming, we react to problems individually, as they become apparent: against weeds, we use herbicides; against insects, insecticides; against deficiency of nitrogen, we apply nitrate fertilizer. Structural change of the farming system may be a better option – looking to cause and effect and the checks and balances that operate within ecological systems. In this transformation, pride of place should be given to the management of soil organic matter. Karlen and Cambardella (1996) have pointed out that the change from crop rotations to monoculture has resulted in a decline in the labile fraction of soil organic matter. The consequences include degradation of soil structure, increased bulk density and reduced permeability – even where the total amount of organic matter remains much the same.

It may be helpful to consider the soil as a bank. We may borrow capital only if there is a commensurate deposit, otherwise we (and the bank) will be bankrupt. We have already lived too long incurring debt for future generations; only now are we beginning to realize the dangers – for us and for those who must follow. Restoration of soil capital is an urgent task for all farmers and agronomists. New policy initiatives are needed to promote the work, and reliable technical measures are needed to monitor progress.

Continuing the theme of borrowing from soil capital, Table 5 (after Katalimova 1960) shows our drawings upon mineral nutrients other than nitrogen in terms of the mineral ash content of various crops.

The ash content of the straw is three times greater than that of the grain and, although made up of much the same elements, the roles of sodium in wheat and calcium in maize are also significant. Sunflower seeds have an ash content of 3.5% with significant contents

**Table 5** Ash content of plants, mean % total mass

| Crop | Ash | K$_2$O | Na$_2$O | CaO | MgO | P$_2$O$_5$ | mg/kg B | Mn | Cu | Zn |
|---|---|---|---|---|---|---|---|---|---|---|
| Wheat, grain | 1.78 | 0.50 | 0.06 | 0.97 | 0.15 | 0.85 | 2.0 | 47.0 | 5.2 | 65.0 |
| Wheat, straw | 4.86 | 0.90 | 0.68 | 0.28 | 0.11 | 0.20 | 4.0 | 60.0 | 1.5 | 16.0 |
| Maize, grain | 1.28 | 0.37 | 0.01 | 0.03 | 0.18 | 0.57 | – | – | – | – |
| Maize, stover | 4.37 | 1.64 | 0.05 | 0.49 | 0.26 | 0.30 | – | – | – | – |
| Sunflower seed | 3.5 | 0.93 | 0.10 | 0.20 | 0.51 | 1.39 | 21.0 | 18.0 | 8.1 | 52.5 |
| Sunflower, whole crop | 10.0 | 5.25 | 0.10 | 1.53 | 0.68 | 0.76 | 72.0 | 47.0 | 3.4 | 25.0 |
| Sugar beet, roots | 0.57 | 0.25 | 0.07 | 0.06 | 0.05 | 0.08 | 17.0 | 50.0 | 6.5 | 17.5 |
| Sugar beet, leaves | 1.42 | 0.50 | 0.10 | 0.17 | 0.11 | 0.10 | 35.0 | 180.0 | 6.9 | 50.0 |
| Tobacco leaves | 14.2 | 5.09 | 0.45 | 5.07 | 1.04 | 0.66 | – | – | – | – |
| Tobacco stems | 7.31 | 1.24 | 0.40 | 1.24 | 0.92 | 0.92 | – | – | – | – |
| Alfalfa hay, at flowering | 5.29 | 1.50 | 0.11 | 2.01 | 0.56 | 0.56 | 68.0 | 86.0 | 6.2 | 25.0 |

of potassium, phosphate, magnesium and calcium; but the ash content of the whole crop is huge – in total 10% of which half is potassium and, also, a substantial amount of calcium. Sugar beet has the lowest ash content and tobacco the greatest – with 14% in the leaves and with equal measures of potassium and calcium. All crops take up large amounts of phosphorus, potassium, calcium and magnesium; also silica, iron and several other micro-nutrient elements. These borrowings should be returned to the soil as fertilizer or manure.

Even if we imagine an optimistic scenario in which half of the amount of nutrients taken up by crops is returned to the field as crop residues and dung, there is still a substantial overdraft on the soil's nutrient capital. Some of the overdraft is made good by the weathering of soil minerals, particularly by the release of potassium by illite clay. However, this is not the case with phosphorus, magnesium, several micro-nutrient elements and, in *Leached chernozem* and *Podzolized chernozem*, with calcium. The dramatic decline in application of phosphate fertilizer in the last 20 years has led to a stripping of phosphorus reserves and, the northern chernozem, probably also of calcium.

The availability of nutrients to plants is not simply a question of nutrient budgets. Supply depends on exchange reactions and the active intermediary role of root and mycorrhizal secretions that may compensate for a nutrient overdraft – but not without limits. That is why ecologically accurate agriculture is so necessary.

# References

Andries SV, V Tiganoc and A Donos 2001 *Bulletin of ecopedologic monitoring (agrochemistry)*. Pontos, Chisinau, 65p (Russian)

Boincean BP 1999 *Ecological farming in the Republic of Moldova (crop rotation and soil organic matter)*. Stiinta, Chisinau, 269p (Russian)

Boincean BP, LT Nica and SS Stadnic 2010 Nitrates leaching from field crops on the Balti steppe. *Academos* 16, 91–98, March

Cambardella CA and ET Elliot 1992 Particulate soil organic matter changes across a grassland-cultivation sequence. *Soil Science Society of America Journal* 56(3), 777–785

Chesniak GI, FI Gavriliuc, IA Krupenikov et al. 1983 The state of chernozem humus. 186–198 in VA Kovda (editor) *Russian chernozem 100 years after Dokuchaev*. Nauka, Moscow

Gliessman SR 2000 *Agroecology. Ecological process in sustainable agriculture*. Lewis Publishers, CRC Press LLC, Boca Raton, FL, 357p

Jenkinson DS and JH Rayner 1977 The turnover of soil organic matter in some of the Rothamsted classical experiments. *Soil Science* 123, 298–305

Karlen DL and CA Cambardella 1996 Conservation strategies for improving soil quality and organic matter storage. 397–420 in MR Carter and BA Stewart (editors) *Structure and organic matter storage in agricultural soils*. Lewis Publishers, CRC Press, Boca Raton, FL

Katalimova MB et al. (editors) 1960 *Guide to mineral fertilizers. Theory and practice of application*. Selihozgiz, Moscow, 551p (Russian)

Parton WI, JWB Stewart and CV Cole 1988 Dynamics of C, N, P and S in grassland soils: a model. *Biogeochemistry* 5, 109–131

Parton WI, DS Ojima and DS Schimel 1996 Models to evaluate soil organic matter storage and dynamics. 421–448 in MR Carter and BA Stewart (editors) *Structure and organic matter storage in agricultural soils*. Lewis Publishers, CRC Press, Boca Raton, FL

Rees RM, BC Ball, CD Campbell and CA Watson 2001 Sustaining soil organic matter. 413–425 in *Sustainable management of soil organic matter*. CABI, New York, NY

# Soil Health and Soil Quality

**Abstract** The unprecedented crop yields achieved by the green revolution depended on heavy usage of fertilizers and agrochemicals, the introduction of responsive crop varieties and application of mechanical power. The non-renewable resources from outside the farming system are no longer cheap, unlooked-for ecological imbalances have arisen as the soil has become a mere substrate for growing crops, and tracts of land have been exposed to erosion. The era of cheap food also saw a flight from the land, comparable to what happened in the Great Depression earlier in the twentieth century. Soil health may be defined as the ability of the soil to function as a living organism – to sustain biological efficiency, water and air quality, and the health of plants, animals and people. This definition recognizes the role of the soil in maintaining food quality and safety but, also, rural livelihoods and the environment. The ecological paradigm is built upon optimal recycling of energy and nutrients by social and biological diversification and on preventing problems rather than dealing with the consequences. For instance, it recognizes that overwhelming outbreaks of weeds, pests and diseases are a consequence of the absence of checks and balances in the farming system that has to be attended to – rather than responding by application of agrochemicals. The same ecological approach may be applied to the fundamental relationships between people and the land, both in terms of land tenure and the relationship between soil health and human health.

**Keywords** Soil health · Soil quality · Public health · Land tenure · Ecological approach

## The Imperative of Change

A new agricultural revolution gathered pace after the Second World War; unprecedented crop yields were achieved through heavy usage of fertilizers and agrochemicals, the introduction of responsive crop varieties and the application of mechanical power. All this depended on non-renewable resources from outside the farming system – above all it depended on cheap oil and gas for energy and for the production of fertilizers and agrochemicals. In this brave new world, the central role of biological processes in soil fertility and the sustainability of farming systems was forgotten.

Unlooked-for ecological imbalances were revealed by the emergence of mad cow disease and its human derivative as a result of putting infected, mechanically recovered animal protein into the food chain. Even now, the consequences of widespread adoption of genetically modified organisms are unknowable. The era of cheap food and new technology saw structural changes in the economy that forced down farm-gate prices for agricultural products and brought a flight from the land comparable to the Great Depression earlier in the twentieth century.

The twenty-first century crisis of farming across eastern Europe, and in many other parts of the world, has concentrated minds on the condition of the soil. Setting aside the sensational and bizarre, there are several direct and uncontested consequences of industrial farming:

– Soil has become a mere substrate for cultivation of crops and fertility is maintained by addition of artificial fertilizers

- Concentration and specialization of production units have separated the previously interdependent roles of crop and animal husbandry
- The brute force of mechanical tillage has replaced the role of soil biota in maintaining a friable tilth
- Freedom from constraints on cultivation laid bare whole landscapes, which resulted in huge losses of soil through erosion

For a long time, these effects were masked by ever-greater application of fertilizers and agrochemicals and by extension of irrigation and the cultivated area. Now, a vicious circle of declining soil health has throttled the increase in farm production that characterized the second half of the twentieth century. When we take account of the energy costs of mineralization of soil organic matter and, especially, the wholesale loss of soil organic matter through soil erosion, the energy expended in farming is greater than the energy harvested in the crops. This is so, even setting aside the expense of transport, processing, etc. – the system is bankrupt. The green revolution has ended with the spectacular global increase in energy costs, and global food security has been eaten away – not only by human mouths but by the drive for renewable energy in the shape of biofuels.

We may now think on one of Kommoner's laws of ecology: "Nothing is for free" (Chernikov and Checheres 2000).[1] Any intervention in natural systems should take account of the functioning of the natural system and rules such as full return of nutrients and energy, rotation of crops and livestock and the limits within which natural resources are substitutable. Instead of harmonizing ecological laws and the market economy (or the command-and-control economy that preceded it in eastern Europe), the market (or the target) took precedence. We are now compelled to recognize that our present farming system is not sustainable. It is aggravating the economic crisis and it is destroying the soil on which we all depend. The ultimate cost of degrading soil health and soil quality will be passed on to future generations.

## An Ecological Alternative

Change we must. The new science of agro-ecology is in the forefront of efforts to increase farm production with lesser inputs of non-renewable resources but its principles go beyond substitution of renewable for non-renewable inputs. Doran and Parkin (1996) define *soil health* as the ability to "function as a living organism within the bounds of the ecosystem and the land holding to sustain biological efficiency, water and air quality, and the health of plants, animals and people." This definition recognizes the role of the soil in maintaining food quality and safety, rural livelihoods and the environment. These are long-term issues but, now that fossil fuels and industrial inputs are no longer cheap, it also makes short-term financial sense to rein back on their use.

The new, ecological paradigm is built upon optimal recycling of energy and nutrients; in natural systems, this is accomplished through social and biological diversification. Ecological agriculture also focuses on preventing problems rather than dealing with the consequences. This means attending to the root causes of problems rather than the symptoms. For instance, overwhelming outbreaks of weeds, pests and diseases are a consequence of the absence of checks and balances in the farming system. They may be contained by applications of risky agrochemicals but the imbalance in the system remains.

Gliessman (1998) and Francis (2002, 2004) argue that we should also take account of the interactions between social and economic issues, not only in primary production but also in distribution and consumption all along the food chain. Certainly, there is no ready recipe for more resilient farming systems; we need to address complex economic, ecological, social and political issues. One of these is the relationship between the size of the economy and the size and system of land holdings. Narrow specialization, cheap power and agrochemicals, mechanization and globalization of the economy have driven ever-increasing farm size and a smaller rural labour force. The new, ecological paradigm depends on more manual skills, for instance, for the integration of animal husbandry with arable crops and application of manure to every field, so it is worth revisiting Cheanov's argument that the haulage costs of rebuilding soil fertility should be one of the criteria for defining the optimum farm size (Cheanov 1924).

---

[1] B. Kommoner's book *Closed Circle* was published in Russian translation by Gidrometeozdat, Leningrad 1974.

Optimum farm size remains an open question in Moldova, which has a rather dense rural population. There is a social dimension bound up with the system of land tenure. The *tragedy of the commons* – where land held in common becomes no-one's responsibility and it does not pay to invest private resources when anybody and everybody can reap the profit – was solved in the West by *freehold*, that is, individual land ownership within a system in which land may be bought and sold like any other commodity. However, in a densely populated country with a market economy, land attracts so high a price that it becomes inaccessible to most people; so the poor can neither buy enough food nor grow their own. Countries that operate an unfettered free market have hardly solved the ecological or social problems of land use, least of all the issue of maintaining the health of the soil.

If good land husbandry and free market economics are really incompatible, then it behoves the state to ensure the maintenance of our common wealth – the soil. Various options are open but, surely, we need an ecological approach to the fundamental relationships between people and the land – whether through establishment of a land trust in which the land remains as private property but is used according to its capabilities as farmland, or land is held cooperatively by rural communities that plan and monitor land use and management, plough back profits to maintain soil quality and diversify production strategies – for instance, through product certification and local processing.

On the issue of soil health and its influence on human health, the father of medicine, Hippocrates, affirmed two and a half thousand years ago that people's health depends on the health of the soil. At the end of the nineteenth century the physician Joséf Fodor, in his book *Hygiene of the Soil* wrote "Both religion and poetry, both practical experience and science, show the human dependence on soil. 'You were born out of the earth and to the earth you will return' teaches the first; 'Mother Earth is our cornucopia' the second; the third recognizes that the soil is the source of food for everyone's well-being; and the last sees it as the regulator of mental development and human health" (Fodor 1893). In the modern era, V.A. Kovda (1990) introduced the concept of the soil's pathology – embracing the complex of aggressive processes of soil degradation that contribute to an imbalance of ecological mechanisms that previously harmonized the activities of the soil. "Soil's pathology is a deep-seated pathology destroying the biosphere". The antithesis of the soil's pathology is the soil's sustainability or resilience in the face of damaging disturbances (Hitrov 2002) that we discussed in section "Soil Sustainability" of chapter "The Soil Cover".

When the farmer followed a horse-drawn plough, he intuitively kept track of the health of the soil. With the advent of mechanization, farmers have lost touch with Mother Earth. In the next chapter, we discuss some practical ways to bring back this close contact between farming and the soil but practical farmers can still make their diagnosis from a spade full of earth – according to its consistency, the abundance of roots and earthworms, and even its scent – and they should!

## References

Cheanov A 1924 *The optimum size of farms, Second edition.* New Village, Moscow (Russian)

Chernikov VA, AI Checheres (editors) 2000 *Agroecology.* Kolos, Moscow, 535p (Russian)

Doran JW and TB Parkin 1996 *Soil health and sustainability. Advances in Agronomy 56.* Academic Press, New York

Francis 2002 Agroecology of water use at the farm and landscape levels. In *Sustainability of water – limited agriculture. Symposium in Cordoba.* European Society of Agronomy and American Society of Agronomy

Francis 2004 Greening of agriculture for long-term sustainability. *Agronomy Journal* 96(5), 1211–1215

Fodor J 1893 *Hygiene des bodens.* Vienna, 322p (German)

Gliessman SR 1998 *Agroecology. Ecological process in sustainable agriculture.* Lewis Publishers, CRC Press LLC, Boca Raton, FL, 357p

Hitrov NB 2002 The notion about soil resilience to external influence. 3–6 in NB Hitrov (editor) *Soil stability to natural and human influences.* Nauka, Moscow 295p (Russian)

Kovda VA 1990 *The pathology of soils and conservation of the Earth's biosphere.* Biological Centre of the Russian Academy of Sciences, Puschino (Russian)

# Farming and Soil Health

**Abstract** Natural ecosystems are characterized by heterogeneity, a balance of energy and nutrients, self-regulation and regeneration of soil form and function. By contrast, conventional farming systems exhibit homogeneity, an imbalance of inputs and outputs, a weak capacity for self-regulation and a lesser degree of adaptation to the constraints of the environment. The building blocks of sustainability for soils and farming systems include the following:

- Matching land use with land capability at the landscape scale and at the field scale
- Crop rotation to combat pests and diseases and to achieve full return to the soil of nutrients and energy – which means optimum application of compost and farmyard manure, and a key role for perennial legumes
- Minimum tillage

With good design and management, rotational systems are more resilient, more productive and more profitable than monocultures. The resilience of ecological systems depends on checks and balances and capitalizes on the natural variability within the landscape. This is not easy to establish in farming systems; it needs to be underpinned by a new holistic science and social organization that embraces various disciplines and practitioners that can learn from one another.

**Keywords** Ecosystems and farming systems · Sustainability · Crop rotation · Compost and manure · Minimum tillage · Holistic science · Social organization

## Natural Ecosystems and Agroecosystems

The market has not resolved the contradiction between financial interests and the laws of ecology; too few farmers are attending to the health of their soil – and this path is fraught with danger. But the accumulating problems of farming systems and our social system may be solved if we take nature as our model. Earlier (Table 1 of chapter "Biological Cycles"), we made a brief comparison of natural and agroecosystems. In more detail, we observe that natural ecosystems are characterized by

- *Heterogeneity* on the one hand and, on the other hand, unity in the interaction of their diverse components within the landscape
- *Balance* – of energy and nutrients, between predator and prey, between parasite and host – all along the trophic chain
- *Self-regulation and regeneration* of soil form and function
- *Interplay* between ecosystems and the social, cultural and economic development of society

In contrast, agroecosystems are characterized by

- *Homogeneity* both at the field and the landscape levels
- *Imbalance* between inputs and outputs of energy and materials; 50–90% of output is carried away from the field as crops and livestock
- *Exploitation* of water and nutrients from a shallow layer of topsoil, exposure of the soil surface during the growing season and vulnerability of crops to pests and diseases

- *Weak capacity for self-regulation and regeneration*, lesser adaptation to environmental constraints and, therefore, dependence upon external inputs to maintain crop production
- *Productivity is determined by the economic and cultural development of society*

In short, agroecosystems will not be sustainable without root-and-branch change at the farm level and landscape level. Many farmers have acknowledged their responsibility for the environment and have introduced alternative, more-sustainable practices – which have also proven financially profitable.

Farmers cannot do much about the soil forming factors of relief, climate and soil parent material but they can influence the course of soil-building processes by creating favourable conditions for plant growth – for instance, by judicious tillage, crop rotation and application of manure and fertilizers. A key feature of sustainable practice is close attention to the management of soil organic matter, in terms of both input and decomposition. Figures 1, 2, 3, 4, 5, 6 and 7 illustrate some aspects of current practice in Moldova and, particularly, Selectia Experimental Station at Balti.

## Principles of Sustainability

### Land Use Planning

Soil and water conservation begins at the landscape scale with allocation of land use according to land capability. We need to establish the optimum ratio and place of arable, meadow, woodland and – especially in the steppe zone which is liable to drought – reservoirs. Sloping land, steeper than 7°, should be withdrawn from arable use; in any case, it is less productive than the plains and continual loss of topsoil under arable usage saps its fertility; however, much of the loss is masked by applications of fertilizer.

The role of woodland shelter belts is equally important to protect fields from searing heat and erosion by the wind and to provide shelter for wildlife and *beetle banks* of useful invertebrates.

### Crop Rotation

Biological diversity may also be conserved by strip cropping, poly-cropping and, especially, by crop

**Fig. 1** "The greatest treasure of Moldavia is the vineyards." *Charles de Jopecourt 1617*

**Fig. 2** Preparation for sowing winter wheat in the Selectia long-term field experiment with different crop rotations

**Fig. 3** Long-term polyfactorial field experiment on the action and interaction of different crop rotations, soil tillage and fertilization

**Fig. 4** Drawing the mouldboard plough at the experimental fields of Selectia Research Institute for Field Crops at Balti

**Fig. 5** Laboratory building of the Selectia Experimental Station at Balti

**Fig. 6** Field day at Selectia Experimental Station introduced by one of the authors, Prof. Boris Boincean

**Fig. 7** Field day on harvesting of winter wheat at Selectia Experimental Station

**Fig. 8** Most senior author, Prof. Igori Krupenikov with his grandson. Photo: Vera Krupenicov

rotation. The more diverse the crops, the greater the soil's biodiversity, and the greater the capacity of the crops to resist pests and diseases.

Not so long ago, it was considered proper for a farm to provide itself with fertilizer. This is only possible by optimizing crop rotation and integrating crops and livestock husbandry – taking account of the combined effect of all the individual components on soil fertility. The proportion of arable crops in the rotation may be high on the plains but it should be less on sloping land – to combat soil erosion. In all cases, the golden rule is to maintain a balance between the input and output of organic matter and nutrients so that the soil's reserves are rebuilt. Convenient procedures have now been devised to assess the carbon and nutrient balance for individual fields and for whole farms – so it is easy to judge whether the farming system is sustainable.

The loss of soil organic matter is much less under crop rotations than under monoculture, and losses are greatest under ploughed fallow which exacerbates the mineralization of organic matter. Legumes, particularly perennial legumes, are an indispensable component of sustainable farming systems but the Norfolk 4-course rotation, with one-quarter of the land under a legume, cannot balance the soil organic matter budget without addition of extra farmyard manure. Our own long-term experiments on chernozem at the Selectia Experimental Station, at Balti, show we are not able to stem the decline of soil organic matter, even with 30% of perennial grasses in the rotation and modest additions of farmyard manure (Table 1, further experimental evidence is provided in the next chapter).

Rotations should be designed to minimize the period during which the soil is bare. Most soil erosion takes place when rain and wind can attack the unprotected soil surface.

It is a pity that we have also forgotten the valuable practice of *catch cropping* – growing green manure crops that bank soil nutrients between cash crops.

**Table 1** Changes in soil organic matter stocks in the 0–20 cm layer under different arable rotations, monocultures, bare fallow and meadow at Selectia Experimental Station, 1962–1991

| Index | Meadow | | Bare fallow | | Continuous cropping | | | | Crop rotations, % row crops | | | |
| --- | --- | --- | --- | --- | --- | --- | --- | --- | --- | --- | --- | --- |
| | | | | | Winter wheat | | Maize, grain | | | | | |
| | Fertilized | Unfertilized | Fertilized | Unfertilized | Fertilized | Unfertilized | Fertilized | Unfertilized | 40 | 50 | 60 | 70 |
| Reduction of stock of soil organic matter relative to the original level | | | | | | | | | | | | |
| t/ha | 6.7 | 15.4 | 30.5 | 37.0 | 17.3 | 22.6 | 19.0 | 22.6 | 13.4 | 16.3 | 14.9 | 14.9 |
| % | 5 | 11 | 23 | 27 | 13 | 16 | 14 | 17 | 10 | 12 | 11 | 11 |
| Annual losses of soil organic matter (t/ha) | – | – | 1.17 | 1.42 | 0.66 | 0.87 | 0.73 | 0.87 | 0.45 | 0.54 | 0.50 | 0.50 |

Neglect of this practice results in eutrophication of groundwater, irrespective of reduced doses of soluble mineral fertilizer. Leaching of nitrates is greatest in the early spring and late autumn when crop water demand is less than the amount of rainfall. During these periods, decomposition of soil organic matter can be a source of nitrates in the groundwater so it is important to know the nitrate potential of each and every soil on the farm and match it with crop demand. This will avoid the application of excess nitrate fertilizer and losses from the mineralization of soil organic matter; nitrate in drinking water is harmful whatever its source.

Now we are seeing a renewed practical interest in correct crop selection, timing of cultivations and sowing (and under-sowing) and alternative ways of pest and weed control – not only to avoid pollution of water supplies but for their financial benefits.

Ecological agriculture nurtures the soil (rather than the plants) by feeding the soil biota with energy-rich materials – not just nutrient salts from fertilizer. From the economic as well as the ecological standpoint, it is better to under-feed crops than over-feed them; surplus nitrate promotes weeds, pests and diseases that must be countered in turn by chemical sprays. It is no accident that the agrochemicals industry prefers to market seeds, fertilizers and agrochemicals (to extirpate the weeds and pests) as a package – and huge sums are expended on studying their interactions.

By contrast, research on ecological agriculture is in its infancy. In Scandinavia, farmers are obliged to transfer 1% of the cost of any agrochemicals they use to a fund for research on ecological agriculture. Without doubt, we need financial support for research, and the results should be widely disseminated, but the main drive to restore soil fertility must come from farmers doing practical experiments in their own fields and exchanging these results in a community of practice. Such an endeavour will need a revival of our lost communication with the soil – so that a new approach to intensified crop production can spring up from below, rather than waiting for it to trickle down from above.

## Compost and Manure

Researches on compost and manure are no less pertinent. When fertilizer is expensive, it pays to make your own by managing the synthesis and decomposition of organic materials outside the soil and then apply the ready-made, well-humified organic matter. Making good manure and compost is a science in itself; just as every housewife, using the same ingredients, cannot make equally tasty bread, not every farmer can make good manure or compost at the first attempt; it is a science but also an art and the more people that engage in it, the better we shall learn from one another.

The cost of carrying forage and spreading muck can be much reduced by sowing rotational grass for grazing livestock. This was a standard practice before the advent of intensive systems of stock feeding. Moreover, dung distributed by livestock in the natural way is fixed more efficiently in the soil than muck spread on the surface and ploughed in. There is also renewed interest in haymaking and silage on pastures as a means of land improvement.

## *Minimum Tillage*

The farmer ploughs to bury weeds and to mobilize nutrients in the soil. These requirements have changed with the advent of herbicides and fertilizers. Many researches, including the long-term experiments at Selectia Experimental Station, demonstrate a continuing role for tillage to control perennial suckering weeds but they, also, reveal the importance of timeliness and quality of tillage – as opposed to its depth and frequency.

A soil of good tilth and permeability, like the chernozem, does not need to be periodically turned upside down. This operation breaks up soil structure; perturbs and slices up the biota; accelerates the mineralization of soil organic matter; and exposes the soil to baking sun, beating rain and drying wind. By contrast, with minimum tillage, the soil surface and the biota may be protected from the elements by a mulch of crop residues. Moreover, frequent small additions of organic matter that can be incorporated into the soil by earthworms are fixed more efficiently in the soil, and losses of nitrogen are much less, than with occasional large applications of manure that have to be ploughed in. An open soil structure can be maintained in a crop rotation by perennial crops with a strong, penetrating root system. But for this to happen we need farm machinery that can accomplish essential operations like sowing on a surface covered with crop residues – and we have to overcome the farmers' ingrained belief in the need to plough and the rightness of the black appearance of the land in autumn and spring.

## Issues of Productivity

When we assess the interaction of crop rotation and tillage, the role and efficiency of fertilizers appear in a different light. Research confirms the efficiency of fertilizers in raising the productivity of monocultures but the yield benefits of crop rotations appear to be greater without fertilizer (Table 2).

The productivity of crop rotations is greater than that of monocultures but the greater response to fertilizer is seen in monocultures. The highest levels of mobile nutrients are also observed in monocultures, and this paradox suggests that the assessment of soil fertility solely on the basis of analysis of instantly available major nutrients does not reflect the whole picture. Simple application of fertilizer is akin to treating a patient with a drug that controls some symptom or other but without taking account of the real cause of the disease, which may be more related to the soul. The soul of the soil is the continuous process of transformation of organic matter. We are far from understanding all of its complex interactions but, at our present level of comprehension, the important thing is to create the right conditions for this process.

## Breaking the Law

With the introduction of artificial fertilizers, it seemed there was no longer any need to take account of unfavourable growing conditions in any particular field or to compensate for any deleterious consequences of one crop with the management of the succeeding crop – the same yield could be achieved simply by applying fertilizer. Actually, the response of a monoculture to fertilizer (or irrigation for that matter) is a pointer to the health of its root system – and the health of the soil. If we dig up the roots of a monoculture, we will see that they are darker than those of a crop grown in rotation (presumably this is connected with the presence of pathogens) and more easily crushed between finger and thumb.

The better rooting systems of rotational crops are better able to draw upon water and nutrients, so the monoculture's response to irrigation and fertilizer is actually a measure of our compensation for the crop's own weak capacity to feed itself. The inevitable decline of fertility under a monoculture demands bigger and bigger doses of fertilizer to maintain yields. We may observe this in long-term trials where yields have stalled in spite of bigger doses of fertilizer and the introduction of high-yielding crop varieties.

Another cogent argument for crop rotation is provided by Goldstein's research at Washington State University (Goldstein 1986); he found that winter wheat competed better against perennial weeds when

**Table 2** Fertilizer efficiency in crop rotations and monocultures in long-term experiments at Selectia Experimental Station, 1991–1996 mean yields, t/ha

| Crops | Crop rotation | | Continuous cropping | | Yield increase in crop rotation | | | Yield increase with fertilizer | |
|---|---|---|---|---|---|---|---|---|---|
| | Fertilized | Unfertilized | Fertilized | Unfertilized | t/ha % | Fertilized | Unfertilized | Crop rotation | Continuous cropping |
| Winter wheat | 4.49 | 4.19 | 3.64 | 1.70 | t/ha | 0.85 | 2.49 | 0.30 | 1.94 |
| | | | | | % | 23 | 147 | 7 | 114 |
| Sugar beet | 39.7 | 32.4 | 33.6 | 14.1 | t/ha | 6.1 | 18.3 | 7.3 | 19.5 |
| | | | | | % | 18 | 130 | 23 | 138 |
| Sunflower | 1.91 | 1.81 | 1.66 | 1.59 | t/ha | 0.25 | 0.22 | 0.10 | 0.07 |
| | | | | | % | 15 | 14 | 6 | 4 |
| Winter barley | 4.03 | 3.68 | 4.23 | 2.54 | t/ha | –0.2 | 1.14 | 0.35 | 1.69 |
| | | | | | % | –5 | 45 | 10 | 67 |
| Maize for grain | 5.12 | 5.44 | 5.32 | 3.74 | t/ha | –0.2 | 1.7 | –0.3 | 1.58 |
| | | | | | % | –4 | 45 | –6 | 42 |

sown after perennial grasses and legumes than following a crop of barley. Soil fumigation, which eliminates pathogens, led to a doubling of the length of wheat roots in a second wheat crop or following barley – but rendered little advantage following 2 years under lucerne. Further observations of crops on farms across the Great Plains under conventional and organic (without chemicals) practice revealed better conditions for root development in rotations that include perennial grasses and legumes and, also, with the inclusion of cover crops and compost. Monoculture promotes root rot and the subsequent compensating development of a dense but shallow root system diverts the plant's energies away from grain production (Goldstein 2000, Goldstein and Boincean 2000).

Again, R.J. Cook's 15 years of research at Rothamsted on the effects of soil fumigation on wheat crops showed that fumigation increased the yield of monocultures by 70%. The beneficial effect of soil fumigation decreased according to the length of the rotation: under alternating wheat and peas, or wheat and bare fallow, fumigation before sowing wheat increased the wheat yield by 22%; in a 3-year rotation of wheat–spring barley–peas, the average benefit due to fumigation was only 7% (Cook and Papendick 1972). We may conclude that an interval under perennial crops is equivalent to soil fumigation and this seems to be the principle benefit of crop rotation (Bezdicek and Granatstein 1989).

Even with the control of above-ground weeds, pests and diseases by agrochemicals, the performance of monocultures is suppressed by unbalanced biological processes within the soil. We come back again to the economy of observing the *Law of Crop Rotation*, as opposed to compensating for its neglect by chemical means. Breaking this fundamental law of agriculture has led to extra expense and various other consequences that are only now becoming apparent: pollution of the environment, risks to public health and the social costs of the flight from the land.

Another law of agriculture, the *Law of Full Return*, requires the return of fertility to the soil, both across the farm and in each individual field; and not only full return of nutrients but also the energy of soil organic matter. It is well known that the power of mineral and organic fertilizers is not equal. Crops produced from mineral fertilizer are not enough to restore the organic matter that is mineralized during the growing season (Smith 1942) so reliance on soil analysis or nutrient budgets to calculate fertilizer requirement is unsatisfactory in the long term. Almost a century ago, Prince Trubecchoi recommended greenhouse and field trials to calibrate the results of soil chemical analysis for practical farming practice (Trubecchoi 1913). We may calculate what is needed to maintain the soil's inherent capacity to supply nutrients by analysis of data from unfertilized strips across the farm. For example, a winter wheat crop producing 5 t of grain per hectare takes up 200 kg/ha of nitrogen. From an application of 60 kg/ha nitrogen fertilizer, the crop may take up, at best, 30 kg – as determined at Selectia Experimental Station (Boincean 1998). The other 170 kg/ha of nitrogen (85% of the crop's requirement) is provided by the soil.

Efficient use of fertilizers is only possible by observing an optimal ratio between natural and artificial (fertilizer) nitrogen. The ratio may be wider in fertile soils and for crops following a generous precursor, compared with poor soils and crops following a hungry precursor. This is the great value of legumes, especially perennial legumes, in a crop rotation; incorporation of legume residues with their favourable carbon:nitrogen ratio improves the soil and the succeeding crop. Re-introduction of perennial legumes into the farming system is indispensable for a sustainable farm economy.

## Issues of Resilience

Ecological farming systems depend on the prevention of pests, diseases and weeds – as opposed to exterminating them. Almost a century ago Pacoschi, a follower of Dokuchaiev who made a close study of the vegetation of Bessarabia, wrote that a combination of practices is needed to control weeds (Pacoschi 1914). If the crop is to compete successfully, it needs a competitive advantage – above all, it needs to be given a good start (Grumer 1957, Ivanov 1973, Tulikov 1982).

Weeds are a good indication of soil fertility; if cultivated plants are overmastered by weeds, then we have not created the right conditions for the crop. Weeds are plants in the wrong place; each has its own niche in nature but they are very adaptable and capitalize on the farmer's mistakes; they appear when conditions are unsuitable for cultivated plants, for instance, in wet soils, acid soils, infertile soils and during drought

(Pfeiffer 1970). Weed seed may always spread from neighbouring areas by the wind, birds or otherwise – but invasion by, say, thistles is also a consequence of a loss of fertility, especially a loss of humus and soil structure; the appearance of bindweed indicates a crusted soil and compaction at the plough sole; and bristle grass (*Setaria* sp.) thrives on compact soils poor in calcium and sulphur.

Balanced nutrition, of itself, enhances the crop's resilience to pests and diseases. Francis Chapeau of the French National Institute for Agronomic Research, in his theory of *trophobiosis*, suggests that pests thrive on crops that are suffering from a disturbed metabolism which raises the level of soluble nutrients in their cells compared with healthy plants – so healthy plants may have a lower nutritional value for the pest (Lutzenbarger 1984). Such an imbalance may arise from unbalanced nutrition, such as excess nitrates. It may also arise from disturbance of metabolism by pesticides, so we need research on the effects of pesticides on crops as well as their effects on pests.

A farming system built upon selection and combination of crops in rotation, with the optimum application of farmyard manure, needs lesser inputs of chemical fertilizer, herbicides and pesticides than does a monoculture. It can create optimal conditions for the transformation and renewal of organic matter, for infiltration and storage of rainfall (which is the main constraint on crop production on the steppes and prairies) and for the arrest of soil erosion. Whereas a mulch of crop residues suppresses weeds and improves the soil's tilth, chemical control of weeds disturbs the ecological balance – suppressing the target pest but favouring another that must then be controlled by further chemical sprays. It is all too easy to control weeds by spraying with herbicides but this does not remove the real cause of their emergence. Instead, ecological agriculture looks to cause and effect and to biological checks and balances.

This is not easy. Transformation of the farming system from feckless exploitation of the soil to sustainable agriculture is one of the great challenges of our time. There is no doubt that the division of scientific enquiry into separate disciplines has enabled in-depth analysis of the separate subjects but it has, also, diverted attention from the holistic study of nature. To meet the challenge of developing holistic, ecological farming systems, science needs to step up to a new level of synthesis and an intellectual organization that embraces various scientific disciplines and active practitioners.

## References

Bezdicek D and D Granatstein 1989 Crop rotation efficiencies and biological diversity in farming systems. *American Journal of Alternative Agriculture* 4, 3, 111–119

Boincean BP 1998a *Crop rotation and restoring soil fertility for arable chernozem soils of Moldova*. Thesis, Dr Hab Agricultural Science, KA Timiriazev Agricultural Academy, Moscow, 304p (Russian)

Boincean BP 1998b *Ecological agriculture in the Republic of Moldova: crop rotation and soil organic matter*. Stiinta, Chisinau, 269p (Russian)

Cook RJ and RI Papendick 1972 Influence of water potential of soils and plants on root diseases. *Annual Review of Phytopathy* 19, 249–374

Goldstein WA 1986 *Alternative crop rotation and management systems for the Palouse* PhD thesis, Dept Agronomy and Soils, Washington State University, Washington

Goldstein WA 2000 The effect of farming systems on the relationship of corn root growth to grain yields. *American Journal of Alternative Agriculture* 15, 3, 101–109

Goldstein WA and BP Boincean 2000 *Farming systems on an ecological basis in the forest-steppe and steppe zones of Moldova, Ukraine and Russia*. Ekoniva, Moscow, 207p

Grumer G 1957 The inter-influence of crops-allelopathy. *Translated from the German by AN Boiarkin*. Foreign Literature, Moscow

Ivanov VP 1973 *Root exudates and their importance for phytocenosis*. Nauka, Moscow, 320p (Russian)

Lutzenbarger JA 1984 *The Ecologist* 14, 2, 1–9

Pacoschi I 1914 Description of the vegetation of Bessarabia. *Reports of the Imperial Society of Agriculture in Southern Russia, Odessa* 5/6, 37–61

Pfeiffer EE 1970 *Weeds and what they tell*. Biodynamic Farming and Gardening Association, Mercury Press, Spring Valley NY, 95p

Smith 1942 *Sanborn Field. Fifty years of field experiments with crop rotation, manure and fertilizers*. Bulletin 458, University of Missouri College of Agriculture Agricultural Experimental Station, Columbia MI

Trubecchoi PP 1913 *Report of the Ploteansk Agricultural Experimental Station Experimental Station for 1912*. The Imperial Society of Agriculture in Southern Russia, Odessa, 380p (Russian)

Tulikov AM 1982 *Weeds and weed control*. Moscow Worker, Moscow, 167p (Russian)

# Experimental Confirmation of the Efficiency of Different Farming Systems

**Abstract** Forty years of data from the long-term field experiments at Selectia Experimental Station at Balti are summarized. The long-term trends of grain yield for Moldova over the same period show a small increase for winter wheat but stagnation of maize yields. In the Selectia long-term trials, absolute yields in crop rotations are significantly greater than for continuous cereals. For continuous wheat, yields have declined steadily on unfertilized plots, but the decline is less on crops receiving substantial dressings of farmyard manure and fertilizer; by contrast, continuous maize responded poorly to fertilizer application. Yields of both wheat and maize grown in rotation have declined on unfertilized plots and, even, declined somewhat with modest dressings of farmyard manure and fertilizer. However, yields have increased steadily under a regime of higher applications of farmyard manure. Polyfactorial trials were established in 1995, involving seven-field rotations with and without perennial legumes, with and without ploughing, without fertilizer and with various combinations of farmyard manure and NPK fertilizer; these trials show that farming practices drive the accumulation or decline of soil organic matter. Soil organic matter status can be raised by crop rotations that include perennial legumes and farmyard manure, especially under minimum tillage. Crop yields have responded positively to this increase on soil organic matter. The use of perennial legumes is significant not only in increasing soil organic matter content but also in reducing the dependence of the farming system on artificial fertilizers, which makes them more sustainable. Because the chernozem has not lost its essential attributes, we can turn around the processes of soil degradation to processes of soil building, principally by maintaining a continual supply of fresh organic material.

**Keywords** Moldova grain yields · Long-term experiments · Crop rotation · Monoculture · Restoration of soil organic matter

## The Fertility of Chernozem

At a meeting to commemorate the 50th anniversary of the Nicolai Dimo Institute, we asked "How can we raise the fertility of our chernozem?" The chernozem has not lost its essential characteristics: the active and resilient adsorption complex, neutral reaction and abundant reserves of plant nutrients; its capacity for self-aggregation that provides mechanical stability, permeability and a high available water capacity; and its unique kind of organic matter. Apart from our own work (Licov et al. 1981, Boincean 1998, 1999) a pantheon of researchers[1] have established that the humification of chernozem is characterized by a quite stable ratio of humic to fulvic acids and, when the more acid sod-podzolic soils are brought into cultivation, their labile humus fraction changes towards that of chernozem. We note, also, a direct correlation between the soil's labile organic fraction and the kind of soil structure that is good for farming.

The data from long-term field experiments at Selectia Experimental Station, discussed in this chapter, demonstrate that the accumulation or decline of soil organic matter is driven by farming practices. This means that decline of the soil's inherent fertility in farming systems is not inevitable: we can turn

---

[1] We may mention Tiurin (1965), Kononova (1951, 1963), Alexandrova (1980), Orlov (1974), Ponamareva and Plotnikova (1980), Likov et al. (2004) and Chernikov (1987).

around the process of soil degradation to the process of soil building by adopting management practices that observe the laws of agriculture and ecology. Foremost amongst these is to feed the soil with an ample and continual supply of fresh organic residues. The sites of the trials are illustrated in Figs. 2, 3 and 4 in the chapter "Farming and Soil Health".

Two important relationships are apparent in the synthesis of 40 years of data from the field trials at Selectia (Figs. 3, 4, 5, 6, 7, 8 and 9):

- The most obvious is the close tie between year-on-year crop yields and cyclical changes in rainfall in the growing season. We cannot change natural laws or natural cycles of drought, but, knowing them, we can mitigate their impacts on farm productivity. Drought is commonly blamed for poor crops, but drought is a fact of life on the steppes and the prairies. It is also a fact that existing farming systems are poorly adapted to this environment. For instance, sowing winter wheat after a late-harvested forerunner does not allow the wheat crop enough time to develop a strong root system and tiller abundantly before the winter sets in. This means that the yield is depressed even in years of good rainfall, let alone in drought years. By contrast, wheat sown after an early-harvested forerunner has the opportunity to establish itself properly before winter.
- We may also note that yields of both wheat and maize have been generally maintained against a background of diminishing inherent soil fertility. Soil degradation has been masked by the increasing efficiency of the farming system in terms of timeliness of operations, higher doses of fertilizer, chemical control of pests and diseases and by the periodic introduction of more productive crop varieties.

## The Mathematical Equations

Table 1 presents the mathematical equations that summarize the trends of grain yields of winter wheat and maize for the period 1962–2002 across Moldova as a whole and for the long-term trials at Selectia Experimental Station.

Figures 1 and 2 present the yields of winter wheat and maize for Moldova as a whole for 1961–2001 and projections to 2011. Under ordinary farm conditions for Moldova as a whole, yields of winter wheat show a substantial increase, with an average annual increase of 24 kg/ha. In contrast, yields of maize have stagnated, even showing a small average annual loss of 3.6 kg/ha.

Figures 3, 4, 5, 6, 7, and 8, show yields from the Selectia long-term field trials of wheat and maize grown as monocultures with and without fertilizer application and in crop rotations. In general, there has been a tendency for increased yields of grains on both fertilized and unfertilized plots, but absolute yields have been greater in the crop rotation as opposed to continuous grains. Rotation N4 receives 12 t of farmyard manure per hectare as well as mineral fertilizers; rotation N2 receives only mineral fertilizer. The percentage shares of the various crops in rotation are

|  | N2 | N4 |
|---|---|---|
| All row crops | 50 | 60 |
| Sugar beet | 0 | 30 |
| Sunflower | 10 | 0 |
| Maize for grain | 40 | 30 |
| Winter cereals | 30 | 30 |
| Peas | 10 | 0 |
| Mixtures of annual cereals and legumes for forage | 0 | 10 |
| Bare fallow | 10 | 0 |

Figures 3 and 4 provide a convincing example of the importance of soil fertility in maintaining wheat yields. Over 30 years, crop rotation N2 (which has one field under bare fallow) has received a modest application of 0.7 t per hectare of farmyard manure, 36 kg N/ha and 30–57 kg P/ha; rotation N4 received 9.2 t of farmyard manure per hectare, 64 kg N/ha and 48–79 kg P/ha. Yields declined in rotation N2 by 31.3 kg/ha/year; by contrast, wheat yields increased by 28.6 kg/ha/year in rotation N4.

Continuous wheat cropping has been carried out with and without application of fertilizers. The fertilized plots received, on average, 11 t/ha of farmyard manure, 70 kg nitrogen/ha, 61 kg phosphorus/ha and 47 kg potassium/ha. Without fertilizer (Fig. 6) yields declined by 8.9 kg/ha/year. Without fertilizer, yields declined by 36.8 kg/ha/year. Comparing these figures with the trends of yields in crop rotations, we may deduce that the inherent

# The Mathematical Equations

**Table 1** Trends of winter wheat and maize yields in Moldova and in long-term field experiments at Selectia Experimental Station, 1962–2002

| Moldova average | In the Selectia long-term experiments | |
|---|---|---|
| | Crop rotations | Continuous cropping |
| *Winter wheat* | | |
| $Y = 24.4 + 0.2409 \times t + 9.20 \times \sin(2 \times pi/40 + 4.80)$ | Crop rotation N2 $Y = 53.4 + (-0.313 \times t + 5 \times \sin(2 \times pi/9 + 4.40))$ | Unfertilized plot $Y = 28.8 + (-0.368 \times t + 2.50 \times \sin(2 \times pi/11 + 2.0))$ |
| | Crop rotation N4 $Y = 39.4 + 0.2857 \times t + 8.70 \times \sin(2 \times pi/45 + 4.9)$ | Fertilized plot $Y = 34.3 + (-0.0898 \times t + 3.0 \times \sin(2 \times pi/11 + 2.30))$ |
| *Maize, grain* | | |
| $Y = 34.7 - 0.360 \times t + 5.0 \sin(2 \times pi/40 + 5.00)$ | Crop rotation N4 $Y = 63.3 (-0.0371 \times t + 9.80 \times \sin(2 \times pi/36 + 4.0))$ | Unfertilized plot $Y = 37.5 + 0.0959 \times t + 8.00 \times \sin(2 \times pi/28 + 0.70)$ |
| | | Fertilized plot $Y = 51.6 + 0.1891 \times t + 3.7 \times \sin(2 \times pi/31 = 4.7)$ |

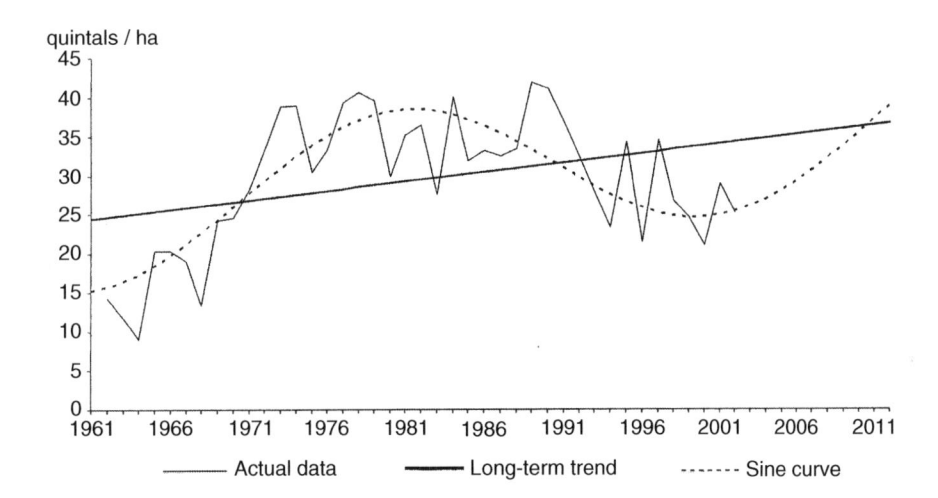

**Fig. 1** Yields of winter wheat in Moldova, 1961–2001, projected to 2011

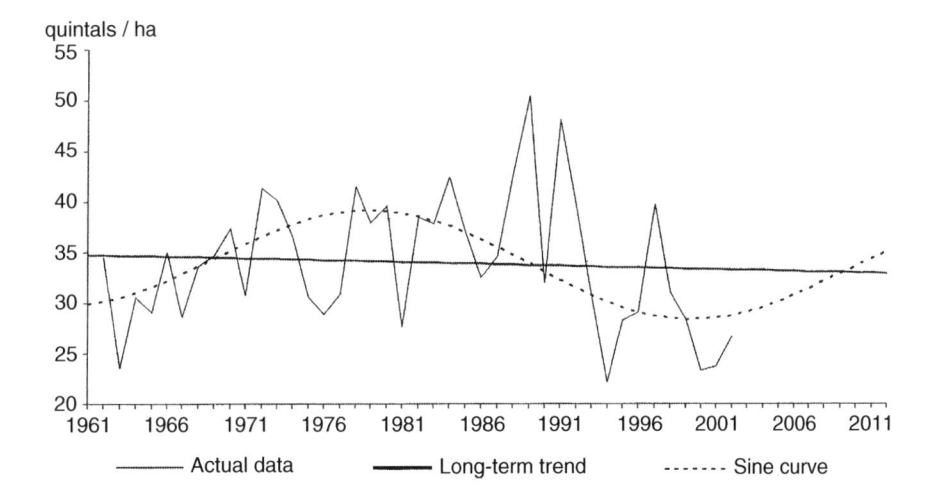

**Fig. 2** Yields of maize in Moldova, 1961–2001, projected to 2011

**Fig. 3** Crop yields in long-term field experiments at Balti, 1962–2002, projected to 2011: Yields of winter wheat at Selectia Experimental Station, rotation 2, 1962–2002. $Y = 53.4 + (-0.3131)*t + 5*Sin(2*Pi/9 + 4.40)$

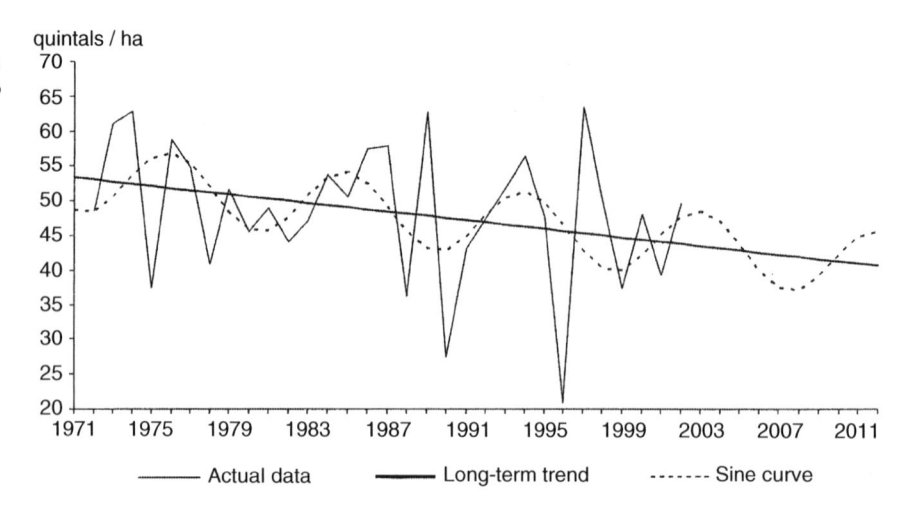

**Fig. 4** Crop yields in long-term field experiments at Balti, 1962–2002, projected to 2011: Yields of winter wheat at Selectia Experimental Station, rotation 4, 1962–2002. $Y = 39.4 + 0.2857*t + 8.70*Sin(2*Pi/45 + 4.90)$

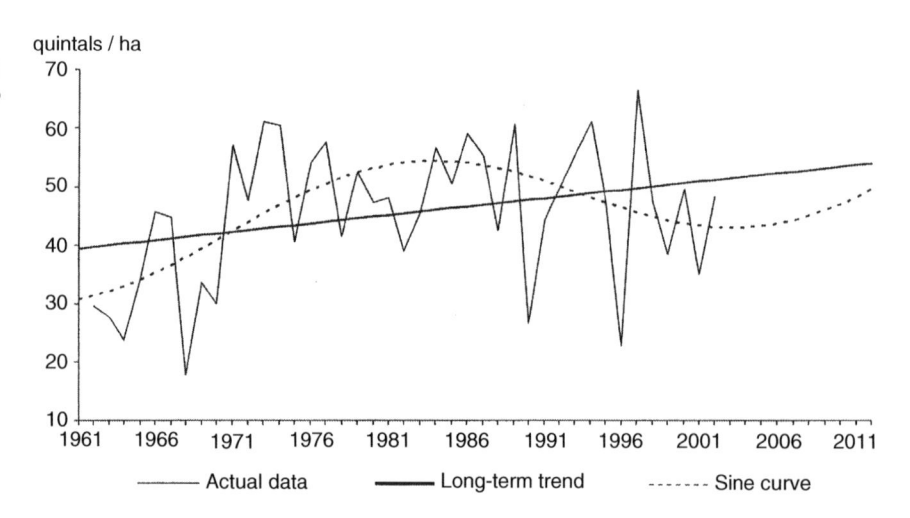

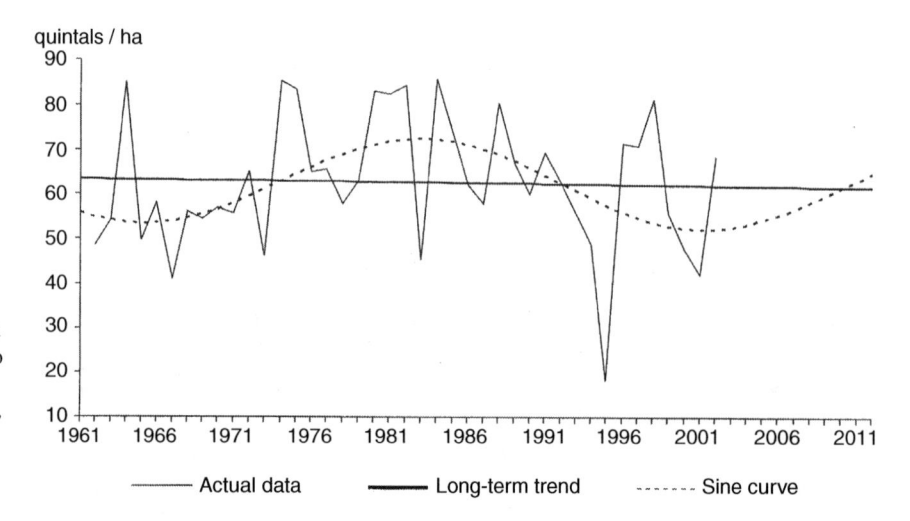

**Fig. 5** Crop yields in long-term field experiments at Balti, 1962–2002, projected to 2011: Yields of maize at Selectia Experimental Station, rotation 4, 1962–2002. $Y = 63.3 + (-0.0371)*t + 9.80*Sin(2*Pi/36 + 4.00)$

The Mathematical Equations 125

**Fig. 6** Crop yields in long-term field experiments at Balti, 1962–2002, projected to 2011: Yields of continuous winter wheat without fertilizer at Selectia Experimental Station, 1965–2002. $Y = 28.80 + (-0.368)*t + 2.550*Sin(2*Pi/11 + 2.000)$

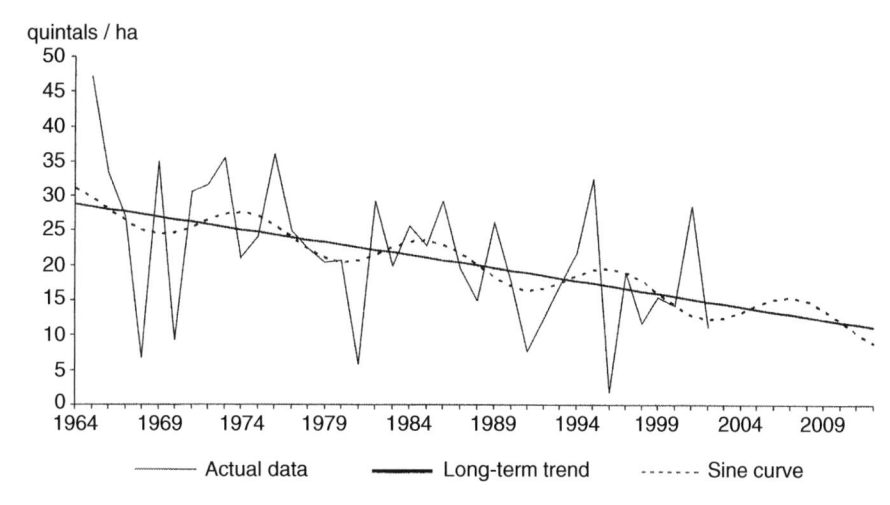

**Fig. 7** Crop yields in long-term field experiments at Balti, 1962–2002, projected to 2011: Yields of continuous winter wheat with fertilizer at Selectia Experimental Station, 1965–2002. $Y = 34.3 + (-0.0898)*t + 3.000*Sin(2*Pi/11 + 2.300)$

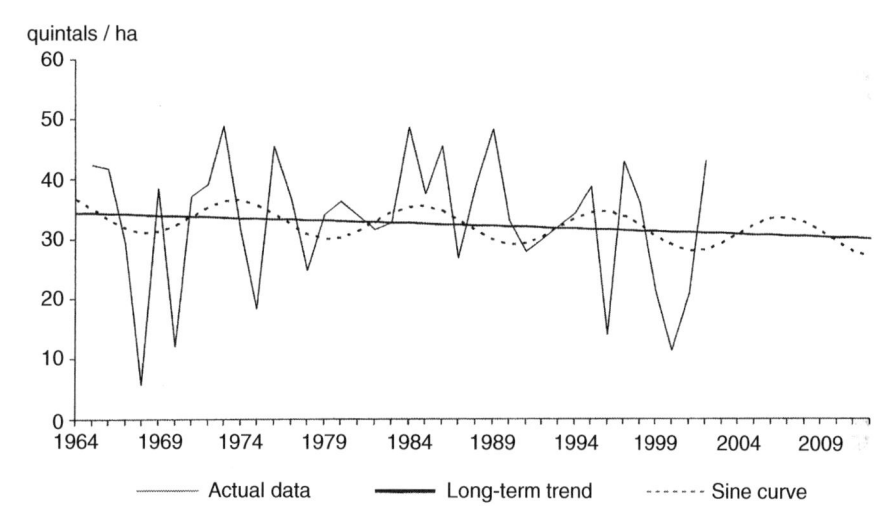

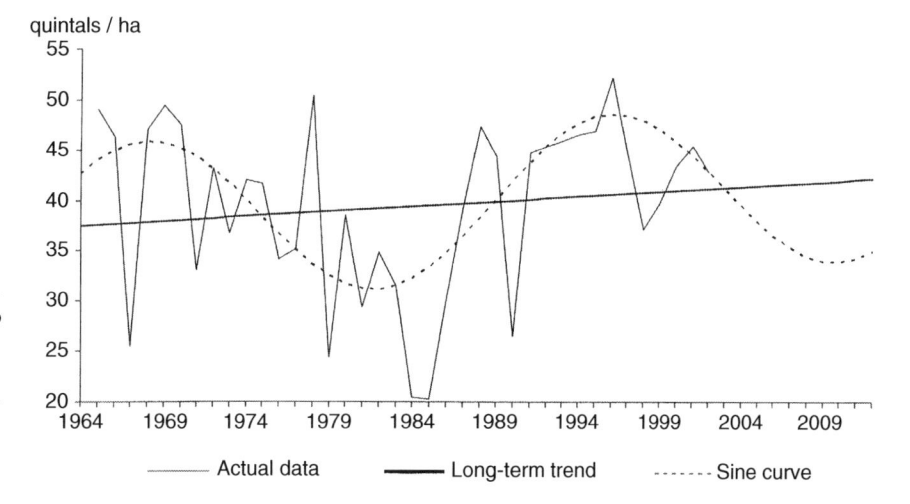

**Fig. 8** Crop yields in long-term field experiments at Balti, 1962–2002, projected to 2011: Yields of continuous maize without fertilizer at Selectia Experimental Station, 1965–2002. $Y = 37.5 + 0.0959*t + 8.00*Sin(2*Pi/28 + 0.700)$

**Fig. 9** Crop yields in long-term field experiments at Balti, 1962–2002, projected to 2011: Yields of continuous maize with fertilizer at Selectia Experimental Station, 1965–2002. $Y = 51.6 + 0.1891*t + 3.70*Sin(2*Pi/31 + 4.70)$

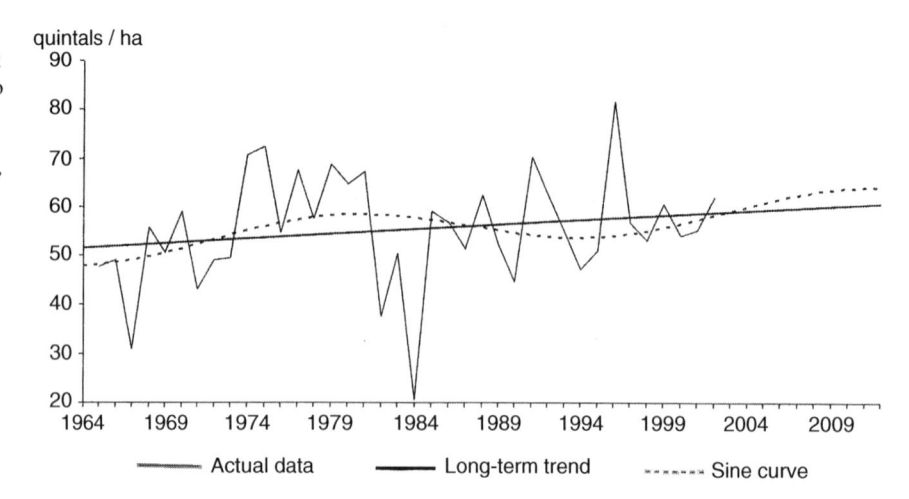

productivity of crop rotations is significantly greater than that of monocultures.

Figures 4, 8, and 9 show a comparable picture for maize. In crop rotation N4 (Fig. 4), maize received the same dressings of manure and fertilizer as the winter wheat crops: 9.2 t/ha of farmyard manure, 64 kg/ha of nitrogen, 79 kg/ha of phosphorous and 49 kg/ha of potassium - and responded with an overall yield decline of 3.7 kg/ha/year. Continuous maize on the fertilized plot received 11 t/ha of farmyard manure, 54 kg N/ha, 80 kg P/ha and 32 kg K/ha - yields on both trials hardly changed, on average rising slightly.

## Changes in Soil Organic Matter Stocks and Crop Yields Under Different Farming Systems

The action and interaction between two crop rotations, two systems of soil tillage and three systems of fertilization are being investigated in another long-term field experiment. Each rotation is 7 years. One includes a mixture of perennial grasses and legumes, and the other does not. One system of tillage is minimum tillage with three passes of the mouldboard plough: to 20–22 cm before sowing maize with under-sown grasses and legumes for silage, to 25–27 cm before winter wheat and to 32–35 cm before sugar beet. In rotation without perennial grasses and legumes, the plough is drawn at 25–27 cm before sowing maize for silage. The other system uses only minimum tillage, without ploughing. Three systems of fertilization include unfertilized plots, farmyard manure and a combination of farmyard manure with NPK fertilizer. The rates of farmyard manure and fertilizer applied annually for the rotation with perennial legumes are 10 t of farmyard manure per hectare and 12.8 kg nitrogen/ha, 21.4 kg phosphorus/ha and 21.4 kg potassium/ha and for the crop rotation without perennial legumes 10 t farmyard manure and 47.1 kg N/ha, 28.5 kg P/ha and 28.5 kg K/ha. No chemical weed, pest and disease control is used.

Each experimental plot is 264 m², and the total experiment occupies 8.7 ha. The full factorial experiment in three replications has been under way since 1995, and two full rotations were completed in 2009.

The soil is *Typical chernozem* with 4–4.5% organic matter[2] in the 0–20 cm layer. Table 2 presents the content of soil organic matter as a mean of three replications for each variant of the experiment. All minimum tillage combinations increased the soil organic matter status of the 0–20 cm layer – between 0.18 and 0.32% over two rotations; for the 20–40 cm layer, there was a slight decrease of organic matter in the plots that received no manure or fertilizer but all other combinations increased the organic matter content between 0.06 and 0.35%. The greatest accumulations of organic matter are in crop rotations

---

[2] Soil organic matter has been determined by wet oxidation according to Tiurin's method.

Table 2 Soil organic matter content (%) in the polyfactorial experiment at Selectia Experimental Station

| Soil layer | Crop rotation with perennial crops | | | | | | Crop rotation without perennial crops | | | | | |
|---|---|---|---|---|---|---|---|---|---|---|---|---|
| | Mouldboard + minimum tillage | | | Minimum tillage | | | Mouldboard + minimum tillage | | | Minimum tillage | | |
| | No fertilizer | Manure | Manure + NPK | No fertilizer | Manure | Manure + NPK | No fertilizer | Manure | Manure + NPK | No fertilizer | Manure | Manure + NPK |
| 0–20 cm | Initial content in 1994 | | | | | | | | | | | |
| | 4.36% | | | | | | 4.30% | | | | | |
| | 1999 | | | | | | | | | | | |
| | 4.24 | 4.14 | 4.51 | 4.13 | 4.32 | 4.41 | 3.95 | 4.10 | 4.05 | 3.96 | 4.13 | 4.20 |
| | 2009 | | | | | | | | | | | |
| | 3.97 | 4.18 | 4.51 | 4.13 | 4.32 | 4.41 | 3.94 | 4.12 | 4.19 | 4.07 | 4.23 | 4.38 |
| | Difference between 2009 and 1999 | | | | | | | | | | | |
| | −0.27 | +0.04 | +0.40 | +0.18 | +0.32 | +0.32 | −0.01 | +0.02 | +0.14 | +0.11 | +0.11 | +0.18 |
| 20–40 cm | Initial content in 1994 | | | | | | | | | | | |
| | 4.07% | | | | | | 4.12% | | | | | |
| | 1999 | | | | | | | | | | | |
| | 3.66 | 3.81 | 3.94 | 3.61 | 3.59 | 3.69 | 3.72 | 3.62 | 3.86 | 3.75 | 3.71 | 3.76 |
| | 2009 | | | | | | | | | | | |
| | 3.32 | 3.82 | 4.36 | 3.82 | 3.94 | 4.04 | 3.51 | 3.73 | 3.98 | 3.51 | 3.90 | 3.82 |
| | Difference between 2009 and 1999 | | | | | | | | | | | |
| | −0.34 | +0.01 | +0.42 | +0.21 | +0.35 | +0.35 | −0.21 | +0.11 | +0.12 | −0.24 | +0.19 | +0.06 |

**Table 3** Yields of winter wheat and sugar beet in the long-term polyfactorial field experiment at Selectia Experimental Station, 1996–2009, t/ha 1 – winter wheat; 2 – sugar beet

| Fertilization systems | Crop rotation with perennial legumes | | | | Crop rotation without perennial legumes | | | | Continuous cropping | | | |
|---|---|---|---|---|---|---|---|---|---|---|---|---|
| | Minimum tillage + ploughing | | Minimum tillage | | Minimum tillage + ploughing | | Minimum tillage | | Minimum tillage + ploughing | | Minimum tillage | |
| | 1 | 2 | 1 | 2 | 1 | 2 | 1 | 2 | 1 | 2 | 1 | 2 |
| Without fertilization | 4.20 | 35.2 | 4.21 | 32.1 | 2.40 | 28.9 | 2.31 | 27.4 | 2.06 | 17.5 | 1.86 | 15.6 |
| Manure | 4.20 | 39.0 | 4.16 | 36.9 | 2.80 | 35.4 | 2.73 | 33.2 | 2.38 | 22.0 | 2.12 | 19.7 |
| Manure + NPK | 4.30 | 40.4 | 4.34 | 38.3 | 3.71 | 40.1 | 3.96 | 36.3 | 2.48 | 22.6 | 2.30 | 20.7 |

with perennial legumes, especially under minimum tillage.

Only the rotation with perennial legumes increased the soil organic matter status under all systems of tillage; the rotation with perennial legumes with the benefit of manure and NPK fertilizer even showed an increase of 0.4% in the 0–20 cm layer in the ploughed plots. In the rotation without perennial legumes, the content of soil organic matter has increased less than the rotation with perennial legumes.

A comparable picture emerges from systematic studies of the influence of agricultural management on soil organic carbon in Canada (VandenBygaart et al. 2003). On their black earths, no-till increased the carbon stocks by 2.9 ± 1.3 t/ha. Carbon storage increased under no-till and reduced summer fallow, crop rotations that include grasses and perennial legumes, ploughing-in green manures and applying fertilizers. The 20-year average soil organic carbon gain factors derived from field experiments and modelling were 0.1–0.14 t/ha/year for no-till, 0.3 t/ha/year for decreasing bare fallow and 0.55–0.56 t C/ha/year for introduction of perennials into the rotation (VandenBygaart and others 2008).

The use of perennial legumes is also valuable for reducing the dependence of the farming system on artificial fertilizers. Table 3 summarizes yield data for winter wheat and sugar beet – two of the crops that are most responsive to fertilization. They show a relatively high level of production in rotation with perennial legumes, irrespective of fertilizer applications. Even on unfertilized plots, the yield of wheat that directly follows lucerne (after the first cut in its third year of life) is close to that of the fertilized plots. The converse is true for the rotation without perennial legumes, in which yields are significantly improved by fertilization.

The trends of productivity, especially for the more exacting crops (winter wheat, sugar beet and forage grasses), indicate the direction of soil building processes. Stagnation or decline of yields should be a warning that the farming system needs to change. In the management of that change, we must look to collaboration of agricultural science with pedology. A key to sustainability is to increase the rate of synthesis of soil organic matter above the rate of expenditure of organic matter and recognizing the dominant role of vegetation in soil building, we must also recognize the irreplaceable role of crop rotations.

# References

Alexandrova LN 1980 *Soil organic matter and the processes of its transformation. Nauka*, Leningrad

Boincean BP 1998 *Crop rotation and recovering the fertility of intensively-farmed arable chernozem soils of Moldova*. Dr Hab Thesis, Agricultural Sciences, KA Timiriazev Agricultural Academy, Moscow, 304p (Russian)

Boincean BP 1999 *Ecological farming in the Republic of Moldova (crop rotation and soil organic matter)*. Stiinta, Chisinau, 269p (Russian)

Chernikov VA 1987 Complex evaluation of the humus state of soils. *Izvestia of the KA Timiryazev Agricultural Academy, Moscow* 6, 83–94

Kononova MM 1951 *Soil organic matter. Its nature, its role in soil formation and in soil fertility*. Academy of Sciences of the USSR, Moscow (English translation by TZ Nowakowski and GA Greenwood, Pergammon Press, Oxford, 1961)

Kononova MM 1963 *Soil organic matter*. Academy of Sciences of the USSR, Moscow (Russian)

Licov AM, VA Chernikov and BP Boincean 1981 Evaluation of humus by characterisation of its labile fraction. *News from*

# References

*the KA Timiriazev Agricultural Academy, Moscow* 1981(5), 65–70

Likov AM, AI Esikov and MN Novikov 2004 *Organic matter in arable soils of the non-chernozem zone. Rosselihozacademia-GNU VNIPTIOU*, Moscow, 630p (Russian)

Orlov DS 1974 *Humic acids in soils.* MV Lomonosov State University, Moscow (Russian)

Ponamareva VV and TA Plotnikova 1980 *Humus and soil (Methods and results of studies).* Nauka (Academy of Sciences of the USSR), Leningrad (Russian)

Tiurin IV 1965 *Soil organic matter and its role in soil fertility.* Nauka, Moscow (Russian)

VandenBygaart AJ, AJ Gregorich and DA Angers 2003 Influence of agricultural management on soil organic carbon: a compendium and analysis of Canadian studies. *Canadian Journal of Soil Science* 83, 363–380

VandenBygaart AJ, BG McConkey, DA Angers, W Smith, H de Gooijer, M Bentham and T Martin 2008 Soil carbon change factors for the Canadian agriculture national greenhouse gas inventory. *Canadian Journal of Soil Science* 88, 671–680

# The Past, Present and Future of the Chernozem

**Abstract** If we consider the history of the chernozem on a time scale comparable to human history, then the pre-Holocene is its prehistoric period, the Holocene its ancient history, the last 5000–10,000 years its modern history and the time from the middle of the nineteenth century up till now the most recent – and cataclysmic – period. The ancient history of the chernozem took place under steppe vegetation; around the northern shores of the Black Sea, sediment laid down and subsequently exposed by the advance and recession of the sea show cycles of chernozem formation with all the features of modern *Calcareous chernozem*. The development of *Common chernozem*, in which carbonates are leached to 0.5 m, clearly takes a long time; the two varieties co-exist in the landscape. Profiles buried beneath Trajan's Bank, about 2000 years ago, represent both *Calcareous chernozem* and *Common chernozem*, testifying to a period of soil formation of 4000–5000 years up to the beginning of the modern period. For 18 centuries after Trajan, farming on the steppe was quite limited. Long grass fallows were practised and historical testaments to the fertility of the country probably reflect the achievement of a stable equilibrium under the contemporary factors of soil formation. But in the most recent period, under the assault of increasingly exploitive agriculture, the resilience of the black earth has been depleted by loss of humus and loss of the topsoil itself by erosion. Four possible futures are considered:

- Back to nature: stopping chemical agriculture and allowing the soil to regenerate itself. This ignores the question of daily bread for a world population of 9 billion by 2050, which will require a doubling of world food production, and the fact that natural regeneration of soils is sluggish
- Business as usual, which will mean erosion of all sloping land within a few decades, widening of the negative humus balance with the balance of available nitrogen and phosphate also becoming sharply negative. It is very likely that this scenario will prevail in the near future. For Moldova this might mean yields falling to 2.1 t/ha of wheat/ha and 2.8 t/ha of maize in 2025 – that is half the level of the 1960s
- A sceptical scenario that aims to arrest rather than reverse the degradation, for instance, by retiring eroded land to forest or pasture
- An ecologically well-balanced scenario to secure enhancement of present levels of fertility and environmental services. This would involve accurate, compensating practices – applying enough farmyard manure, the best possible return of straw to the soil, grass leys, minimum tillage and addition of fertilizers. The results of the long-term experiments at Selectia Experimental Station show how well the chernozem can respond. The aim will be a chernozem lower in humus than the pristine steppe soil but, all the same, a resilient chernozem with its favourable organic matter, adsorption complex and neutral reaction, and rich in nutrient reserves

**Keywords** Soil evolution · Soil history · Archaeology · Future scenarios

## Curriculum Vitae

The chernozem is a phenomenon in the world of soils and amongst Earth systems as a whole; for

Moldova, it is the chief natural asset. As in any soil, the mineral (geological) and living (biological) components are complementary. The mineral part ranks highly amongst soils, providing a stable and potentially valuable inheritance but, in itself, it is of no agricultural value. The biological part that makes the chernozem fertile is unstable and, at present, much perturbed; its future is in our hands. The first task in creating a sustainable farming system is to restore the harmony between the geological and biological components of the soil.

Previous chapters have related different facets of the chernozem's history and present status – from which we may guess at various possible futures. In this chapter we draw together this information in the shape of a biography of the chernozem or, rather, its *curriculum vitae* – for it tells its own story and we are merely re-telling it, drawing on scientific studies. The present is shown by a welter of measured data on various soil characteristics. The future can only be presumptive, in the shape of forecasts. Society's imperative is to choose a future that is ecologically sustainable and, at the same time, pragmatically feasible.

If we consider the history of the chernozem on a time scale comparable to human history, then the pre-Holocene period is its prehistoric period; the Holocene its formative ancient history; the last 5000–10,000 years its modern history; and the time from the middle of the nineteenth century until now the most recent – and cataclysmic – period. In this history, we may distinguish different regimes in the country south of Codri, the uplands of Codri itself and the country to the north; each of these regions has a different landscape and geological history. We will set aside Codri as an ancient place where the chernozem is met with only in the river valleys. Mainly, we will consider the country to the south, for which we have much more evidence of soil evolution than we have from the north.

Paleogeographers like Adamenko (1989), using evidence of the Quaternary era from geochemistry and pollen analysis, conclude that during the period of human occupation "The landscapes of southern Moldavia were dominated by dry steppe ... forests had a restricted distribution" (Alexandrovski 1983). North of Constanta, at the southern extent of the Danube delta, we may observe some 20 m thickness of sediments laid down under periodic advances and retreats of the Black Sea. These sediments have been studied by Ana Conea (1970) and Fig. 1 is drawn from her data.

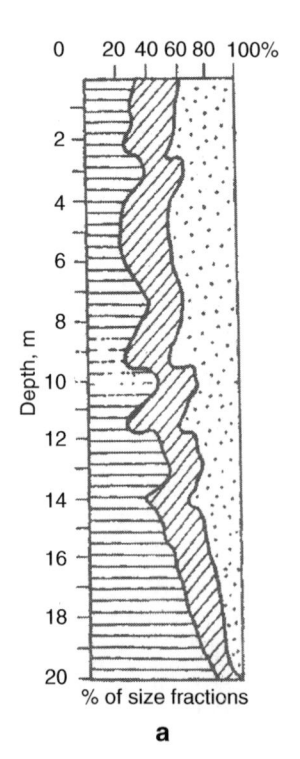

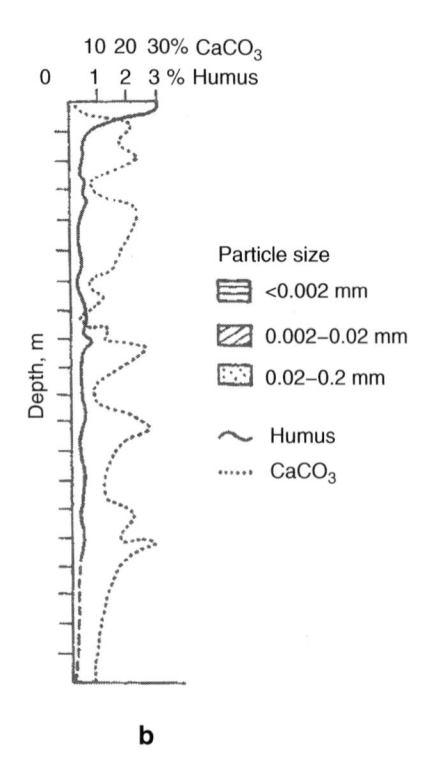

**Fig. 1** Distribution of humus, carbonates and clay in sediments exposed by the retreat of the Black Sea at Constanza; 1a Particle-size distribution; 1b Carbonates and humus

The section depicted in Fig. 1 shows, at the surface, the modern soil – a *Calcareous chernozem* with carbonates throughout the profile and 3% of humus in the topsoil. There are four *paleosols* buried at depths of 3–4, 8–9, 10–11 and 13–14 m. These buried soils are marked by lesser peak contents of humus than the modern soil but also an increase in the content of clay (particles finer than 0.002 mm) which is evidence of the process of clay synthesis that is characteristic of the *Calcareous chernozem* and was first revealed by Screabina's particle size analyses (1972), as discussed in chapter "Soil Texture and Structure", and confirmed by Alekseev's analysis of clay mineralogy (1999), as discussed in chapter "Soil Mineralogy and Elemental Composition". Below each black earth layer there is an accumulation of carbonates.

In the country around Constanta, each period of chernozem formation was interrupted by further sedimentation and, then, resumed in the same way as happens now. The present processes of soil formation may be observed in thick soil profiles across the watersheds of the rivers Bic and Botna near Chisinau that show black earth topsoils with humus accumulation accompanied by clay formation, with carbonate concentrations below, crotovinas . . . In other words, all the attributes of *Calcareous chernozem*.

From these data we conclude that the process of chernozem formation, under the influence of steppe vegetation, has remained remarkably constant and the process has its own momentum. The first soil to form could only be the *Calcareous chernozem* – because the soil parent material in this part of the world has a mean carbonate content of about 16%. We presume that the formation of *Common chernozem*, in which carbonates are leached to a depth of 0.5 m, must take a long time. Both *Calcareous* and *Common chernozem* co-exist in the landscape and the climatic fluctuations over the last 8000–10,000 years, being reversible (Genadiev 1990), have played no great role in the present distribution of the two sub-types of chernozem.

For the region of *Typical chernozem*, in the north of the country, we do not have evidence of buried soils. Today's *Typical chernozem* must have passed through a *Calcareous chernozem* phase because the loess parent material contains, on average, 18% of calcium carbonate. Nowadays, carbonates only occur below 0.7 m depth, which tells a story of gradual leaching under the more humid conditions of the north of the country. From time to time, the more humid parts of the steppe have been invaded by forest – but the essential character of the soil has been renewed. Again, we may conclude that the chernozem is a self-renewable resource and we should be able to drive this process by good farming.

Recently, V.E. Alekseev (2003) has revisited the question of the age of soils and the period of the soil-forming processes. He points out that the period of chernozem formation extends far back into soil history but the age of individual buried profiles and, especially, the modern one is much shorter.

Moving on to the modern period, we can compare today's soils with the soil profiles buried some 2000 years ago beneath Trajan's Bank[1] that we first described in 1960 and, again, in the report on the Moldavian chernozem (Krupenikov 1967; Fig. 2).

Comparing the modern and buried soils:

- The humus and carbonate profiles of the modern soil and buried soils are much the same. The soil developed on the bank itself is also similar to the modern soil except for a humus peak at a depth of 1 m that we attribute to a layer of turfs laid down by the builders to lend the bank stability
- The modern and buried soils have the same humus form with identical humic acid:fulvic acid ratios but the younger profile developed on the bank itself has a lower humic acid content (Table 1)
- The buried profiles represent both *Calcareous chernozem* and *Common chernozem*, testifying to a period of soil formation of 4000–5000 years up to the beginning of the modern period
- The buried profiles exhibit an increased clay content compared with the parent material (30.9–32.5% compared with 20%) and a well-developed, water-stable structure (67% aggregates > 0.25 mm size and 47% > 1 mm size compared with 24 and 2%, respectively) in the parent material (Table 2).

For 18 centuries after Trajan, cultivation on the steppe was quite restricted; long grass fallows were practised and the slopes were hardly cultivated. The

---

[1] Trajan's Bank comprises two sets of earthworks that extend across the south of the country marking the furthest extent of the Roman Empire in the reign of the Emperor Trajan early in the second century AD.

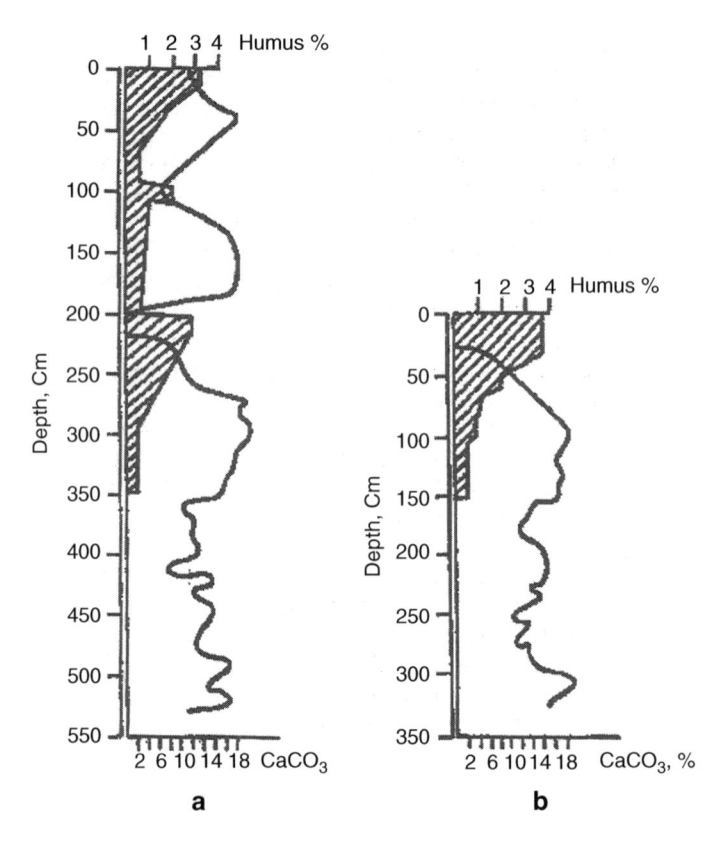

**Fig. 2** Humus and carbonate profiles of Trajan's Bank. 2a Trajan's Bank and buried profile; 2b Control natural soil profile

**Table 1** Structure of humus in the Trajan's Bank soils

| Depth (cm) | Organic carbon (%) | Carbon groups, % of total organic carbon | | | Humic acid/fulvic acid |
|---|---|---|---|---|---|
| | | Bitumen | Humic acid | Fulvic acid | |
| *Soil on the bank* | | | | | |
| 0–9 | 2.18 | 7.1 | 23.4 | 15.0 | 1.5 |
| 30–40 | 0.85 | 3.4 | 32.8 | 16.8 | 1.9 |
| *Buried soil* | | | | | |
| 200–210 | 1.58 | 4.1 | 53.4 | 15.8 | 3.3 |
| 230–240 | 1.11 | 5.3 | 41.8 | 19.4 | 2.2 |
| *Control profile, modern soil* | | | | | |
| 0–20 | 2.18 | 1.7 | 50.1 | 15.2 | 3.2 |
| 40–50 | 1.17 | 2.5 | 49.0 | 20.8 | 2.3 |

historical testaments to the fertility of the region that we relate in section "A Short History" of chapter "Introduction" probably reflect the achievement of a stable equilibrium of humus content under the contemporary factors of soil formation. But in the late nineteenth century, just as the chernozem was attracting the first attentions of science from Grosul-Tolstoy and Dokuchaev, it fell prey to an increasingly exploitive agriculture. For the last century and a half it has maintained an extraordinary fertility but, over wide areas, its resilience has been depleted by loss of humus and loss of the topsoil itself by erosion.

**Table 2** Aggregate-size distribution in the Trajan's Bank soil and buried profile 250 K

| Depth (cm) | Content of fractions (mm) (%) | | | | | | |
|---|---|---|---|---|---|---|---|
| | >0.1 | 0.1–0.05 | 0.05–0.01 | 0.01–0.005 | 0.005–0.001 | <0.001 | <0.01 |
| 0–10 | 0.7 | 11.1 | 49.7 | 2.7 | 13.4 | 22.3 | 38.4 |
| 10–20 | 0.7 | 9.1 | 53.2 | 6.1 | 9.5 | 21.4 | 37.0 |
| 20–30 | 0.4 | 10.3 | 50.2 | 6.1 | 13.6 | 19.4 | 39.1 |
| 30–40 | 0.4 | 14.0 | 41.2 | 11.1 | 17.5 | 15.8 | 44.4 |
| 50–60 | 0.2 | 8.0 | 46.7 | 9.0 | 11.4 | 24.7 | 45.1 |
| 95–105 | 0.3 | 5.9 | 47.3 | 15.7 | 2.5 | 28.3 | 46.5 |
| 130–140 | 0.3 | 9.3 | 48.2 | 2.9 | 15.4 | 23.9 | 42.2 |
| 180–190 | 0.4 | 9.0 | 41.3 | 11.3 | 13.8 | 24.2 | 49.3 |
| 200–210 | 0.4 | 10.3 | 40.3 | 7.3 | 9.2 | 32.5 | 49.0 |
| 220–230 | 0.5 | 7.3 | 48.3 | 6.8 | 6.2 | 30.9 | 43.9 |
| 240–250 | 0.5 | 4.2 | 47.4 | 12.8 | 9.4 | 25.7 | 47.9 |
| 290–300 | 0.0 | 7.4 | 51.7 | 5.8 | 13.1 | 22.0 | 40.9 |
| 340–350 | 0.0 | 13.5 | 51.2 | 11.3 | 7.4 | 16.6 | 35.3 |
| 390–400 | 0.0 | 13.1 | 46.6 | 8.9 | 13.5 | 17.9 | 40.3 |

Whereas the full-profile chernozem is the finest soil in the world for farming, severely-eroded chernozem is a world away. In some places, its unique attributes and even the chernozem soil-forming process are now lost.

New ideas about the genesis of the chernozem connect its past, present and future. The old idea, preserved in Alekseev's treatise on mineralogy (1999), looked on the different sub-types as a linear series from the *Calcareous* to the *Podzolized chernozem*. However, evolution of all the characteristic features of the chernozem does not proceed in phase. By common consent, the *Typical chernozem* is the apogee of fertility and ecological function but, in terms of mineralogy, it is already degraded; in comparison with the *Common chernozem*, it has lost some 12.5% of the clay minerals – yet has 20% greater productivity. Again, if we compare the *Calcareous* and *Common chernozem* with the *Xerophyte-wooded chernozem,* the last is degraded in terms of clay minerals but it is twice as active in terms of humus content, soil structure, permeability, biological diversity and activity (Krupenikov 1967).

Rather than considering the development of chernozem as a single, linear sequence, we propose two separate evolutionary pathways: the mineralogical and the humus accumulating. The first is a long and very slow path; it is irreversible and close to the geological although it can be renewed by a new phase of sedimentation or mountain building. The second is reversible – the amount of humus can increase or decrease according to shifts in the balance of soil-forming factors or a change in the farming system; the ecological vulnerability of the chernozem but, also, its capacity for renewal or even increase in productivity is bound up with this pathway. In these two pathways of evolution we see the linking of two systems: the big, slow, geological system and the smaller, faster, biological system – which harks back to Viliams.

The condition of the chernozem in the 1960s and 1970s was recorded quantitatively in our book *Statistical Parameters of Structure and Properties of Moldavian Soils* (Krupenikov 1981). Only recently have we perceived, within the mass of data, yet another paradox: the fertility of chernozem is frequently defined by the thickness of black earth (with a humus content equal or greater than 1%) and yet the average thickness of this layer is the same in full-profiled members of each sub-type, from *Calcareous* to *Leached chernozem* (100–110 cm, $n = 166$). It is as though the chernozem do not need more than 110 cm to accomplish its functions in the ecosystem. Even more remarkably, the thickness of the topsoil or A horizon is also the same for all sub-types, between 45 and 48 cm ($n = 2177$ descriptions by not less than 60–70 individual pedologists). In spite of all the differentiations based on the slow evolutionary path of mineralogy and leaching of carbonates and the fast track of humus accumulation, it seems that a thickness of 45–48 cm is quite enough for the functioning of the topsoil.

The A horizon, which is diagnostic for the field identification of chernozem, determines the physical and biological function of two-thirds of the soil as a whole (Dobrovolsky and Nikitin 1990). Across most of western Europe, the topsoil has been created by agricultural management – we might say that it has been cloned. Under present management, the arable soils of western Europe are more productive than the chernozem but they lose their fertility much more easily if they are not continually maintained. For this reason, the outstanding German expert K. Mengel (1981) considers that the chernozem should be held up as the gold standard for the assessment and understanding of soil fertility. By way of illustration, the average yield of corn for the 25-year period 1965–1990 in Moldova was 4.36 t/ha on *Calcareous chernozem* and 4.37 t/ha on *Common chernozem*. Application of fertilizer (P60, NP60 and NPK60) made no difference to the maximum yield of *Calcareous chernozem* at 4.48 t/ha; on *Common chernozem* NP60 and NPK60 fertilizer raised the yield to 6.03 t/ha but even yields on control plots remained close to the soil's *bonitet* rating – such is the quality of the chernozem!

## Future Scenarios

One possible version of the future began about 30 years ago with the acceleration of soil erosion and the continued loss of soil organic matter and nutrients. Over this period, the application of artificial fertilizer has decreased 20-fold and application of farmyard manure all but ceased.

Nearly 20 years ago (Krupenikov 1992), we described four possible future scenarios that remain valid today:

– The nostalgic, *back-to-nature* idea which gained momentum in the 1990s. In short: leave nature, including the soil, in peace; stop chemical agriculture and nature will regenerate itself. This ignores the question of daily bread for a world population of 9 billion by 2050, which will require a doubling of world food production to meet the Millennium Goals, and also the fact that natural regeneration of soils is a sluggish business. However, it raised unrealistic hopes in some farmers and, psychologically, opened the way for the organic farming movement
– The *pessimistic* scenario envisages continuation of the recent past, which will mean erosion of all sloping land within a few decades. Already in Moldova some 200,000 ha are severely eroded – their topsoil stripped away; they are not chernozem anymore. A further 250,000 ha are moderately eroded – they have lost many of the attributes of chernozem and the annual negative humus balance has widened to –0.4 t/ha, with the balance of available nitrogen and phosphate also becoming sharply negative. It is very likely that this scenario will prevail in the near future

Leading agronomists have expounded on the future of our soils based on the changes in humus and nitrogen content of the topsoil since 1897 and, especially, since 1965 (Andries 1999, Andries and Zagorcea 2002). Their figures for the humus and nitrogen content of the top 30 cm of Moldovan arable soils in 1965 are 130 and 6 t/ha, respectively, yielding 3.2 t/ha of winter wheat and 4.2 t/ha of maize. For 1996, the humus and nitrogen values were 110 and 5 t/ha of humus and nitrogen, respectively, yielding 2.5 t/ha of winter wheat and 3.4 t/ha of maize. Their forecast for 2025 is 90 t of humus/ha and 4 t of nitrogen/ha, yielding 2.1 t of wheat/ha and 2.8 t of maize/ha. These are serious people and we should take their forecasts seriously, although they are pessimistic in not taking account of the nutrient reserves in the whole soil profile.

– The *sceptical* scenario, aims to arrest rather than reverse the degradation, for instance by retiring eroded land to forest or pasture. Certainly, it would be good to stop further erosion but even this limited approach has not been undertaken
– The *optimistic* scenario, to secure enhancement of present levels of fertility and environmental services, is not real at any scale at present

On the basis of the hard scientific evidence of long-term trials at reputable experimental stations, and synthesis of a mass of data on good farm practice, we suggest a cautiously optimistic scenario that may be dubbed *ecologically well balanced*. This would involve accurate, compensating practices – applying enough farmyard manure, the best possible return of straw to the soil, grass leys, minimum tillage and addition of fertilizers – by which means fertile soils have been created in several countries of western Europe (even from *podzols*, *luvisols* that are naturally depleted of clay in the topsoil, and drained *gley* soils that all began with an unfavourable kind of humus, a world away from our chernozem).

If we introduce such compensatory agricultural practices on the chernozem, then the potential for improvement will be better than in western Europe. The results of the long-term experiments at Selectia Experimental Station, especially the high crop yields and buildup of soil organic matter under best practice reported in section "Changes in Soil Organic Matter Stocks and Crop Yields Under Different Farming Systems" of chapter "Experimental Confirmation of the Efficiency of Various Farming Systems", show how well the chernozem can respond. The aim will be a chernozem lower in humus than the pristine steppe soil but, all the same, a resilient chernozem with humate-dominated organic matter, favourable adsorption complex and neutral reaction, rich in mineral reserves for release of nutrients. Certainly, an increase in applications of farmyard manure is essential and we would do well to heed the appeal of the Roman agriculturalist Varrona in the first century AD on "the great union of agriculture and stockbreeding". He put dung foremost amongst animal products – above meat and milk; maybe an extreme position but a great insight.

## Conclusion

The Brundtland Commission[2] defined sustainable development as "development that meets the needs of the present without compromising the ability of future generations to meet their own needs." Keeping the soil productive and in good health is a key issue in sustainable development. It is obvious that this is not being achieved by farming today. We cannot continue much longer along the path of intensification that demands unlimited use of limited natural resources – with its land degradation, pollution of soil and water and destruction of both biodiversity and rural communities. What is more, the costs of restoring the environment are being passed on to future generations.

The search for an alternative, sustainable farm economy attracts farmers and scientists from all parts of the world but scientists and practitioners differ about how to achieve sustainability. Without necessarily arguing that artificial fertilizers should be excluded, most agree that we must reduce dependence on non-renewable resources. At the same time, globalization and liberalization of world trade seems to be driving farming down the industrial route – with all its damaging consequences.

The ecologically balanced scenario would mean a change in the whole structure of the agricultural economy, a different balance of imports and exports of food and fertilizer, more organically grown food for the home and export markets, new employment opportunities and stabilization of the soil cover. The script is not utopian: it is quite feasible as a combination of efforts by national governments, farmers, local communities and civil society organizations – but it is essential that they operate at an optimum scale for the symbiosis of arable and livestock husbandry. The vision, the explanation and propagation of good farming are necessary but, without financial incentives and due legal basis, it will not happen; we may draw a parallel with the slow development of renewable energy resources. The necessary legal basis is an unambiguous law on soil and soil management that includes consolidation of land holdings, avoiding gigantism but also avoiding splintered and uneconomic smallholdings.

Looking at the issue from the perspective of soil as a living organism that lives by continuous recycling of soil organic matter, this dynamic can be taken as the starting point in a step-by-step modernization of farming systems. The aim of ecological agriculture is to create more favourable conditions for this vital activity and, thereby, a greater variety of living organisms in the soil and a more resilient and more productive farming system. Its foundation is the management of soil quality and crop and soil health. This depends on good husbandry of the landscape as a whole and of each of its inter-dependent parts: soils, crops, farms, rural communities. It needs to include a landscape approach to land tenure and land use planning that can match the various enterprises with those facets of the landscape that can best suited to them; the symbiosis of arable and livestock husbandry; and a sufficient balance of soil organic matter within the limits of each crop rotation, each field and across the farm and landscape as a whole.

This would be a transformation from speculative pillaging of the soil to a husbandry that fully returns what is taken out. Achievement of sustainability along the entire food chain, from production to consumption and return, will not only maintain the soil but go

---

[2] World Commission on Environment and Development (1987).

some way to rebuilding a resilient rural sector – and renewing society.

## References

Adamenko OM, SI Medianic and NF Mirlean 1989 Climate reconstruction in the south of Moldova using the methods of Palaeontology and geochemistry. 66–73 in *palaeogeography and lithology*. Stiinta, Chisinau (Romanian)

Alexandrovski AL 1983 *The evolution of the soils of the Western European Plain in the Holocene*. Nauka, Moscow 150p (Russian)

Alekseev VE 1999 *Mineralogy of soil formation in steppe and forest-steppe zones of Moldova: diagnosis, parameters, factors, processes*. Stiinta, Chisinau, 240p (Russian)

Alekseev VE 2003 On the age of soils and soil formation in the territory of Moldova. 41–51 in SV Andries (editor) *Soil – one of the main problems of the 21st Century. International scientific and practical conference on the 50th anniversary of the Nicolai Dimo Research Institute for Soil Science and Agrochemistry*. Pontos, Chisinau (Romanian)

Andries SV and Zagorcea CL 2002 Soil fertility and agricultural service in agriculture, *Bulletin of the Academy of Sciences of Moldova: Biological, Chemical and Agricultural Sciences, Chisinau* 2, 42–44

Conea A 1970 *Loesses and paleosols in Dobrojea*. Ceres, Bucharest, 234p (Romanian)

Dobrovolsky GV and ED Nikitin 1990 Soil functions in biosphere and ecosystems. Nauka, Moscow, 253p (Russian)

Genadiev AN 1990 *Soils and time: models for development*. Nauka, Moscow 220p (Russian)

Krupenikov IA 1967 *Chernozem of Moldova*. Cartea Moldoveneasca, Chisinau, 427p (Russian)

Krupenikov IA (editor) 1981 *Statistical parameters for soil composition and properties in Moldova Vol. 2*. Stiinta, Chisinau, (Russian) 253p

Krupenikov IA 1992 *Soil cover of Moldova*. Stiinta, Chisinau, 264p (Russian)

Krupenikov IA 1960 Soils down Trajan's Bank and some issues of palaeopedology. 55-69 in *Nature Conservation of Moldova. Vol. 1*, Stiinta, Chisinau (Russian)

Mengel K 1981 Wesentliche factoren der bodenfruchtbarkeit. *Bodenkultur* 32, 9, 189–194 (German)

Screabina IE 1972 *The process of clayization in chernozem soils of Moldova* (short version of dissertation) MV Lomonosov State University, Moscow, 28p (Russian)

WCED 1987 *Our common future*. World Commission on Environment and Development. Oxford University Press Oxford

# Author Index

**A**
Alekseev, V. E., 14, 27–28, 35, 78, 133, 135
Andries, S. V., 18, 43, 101, 136
Atamaniuk, A. K., 70, 73

**B**
Bai, Z. G., 94
Barasnina, L. N., 63, 70
Berezin, P. N., 69, 71
Bezdicek, D., 118
Boincean, B. P., 7, 63, 100, 102, 113, 118, 121
Bondarev, A. G., 4, 70
Burlacu, I., 43, 58

**C**
Cheanov, A., 106
Chendev, I. G., 78, 91
Chiricov, V. F., 58
Cook, R. J., 118

**D**
Daradur, M., 73
Darwin, C., 65
Dent, D. L., xiv, 12–13
Dimo, N., xiii, 6–7, 29, 52, 65, 121
Dobrovolsky, G. V., 136
Dokuchaev, V. V., 4–7, 15, 27, 43–44, 72–73, 86, 134
Doran, J. W., 106

**F**
Fokin, A. D., 33
Francis, C. A., 106, 119

**G**
Ganenco, V. P., 45
Genadiev, A. N., 133
Glazovskaya, M. A., 40
Gliessman, S. R., 100, 106
Goldstein, W. A., 117–118
Gradusov, B. P., 24, 29
Grati, V. P., 46
Grumer, G., 118

**H**
Hitrov, N. B., 17, 107

**I**
Ivanov, V. P., 118

**J**
Jacks, G. V., 70
Jenkinson, D. S., 99
Jenny, H., 15

**K**
Karlen, D. L., 102
Korduneanu, P. N., 43, 63
Kostichev, P. A., 7, 73
Kovda, V. A., 44, 107
Kovali, T. A., 74
Krupenikov, I. A., 5, 7, 14–15, 17, 20–21, 28–29, 34–35, 42–43, 48, 52–53, 55, 58, 60, 63–64, 70, 73, 78, 84, 86, 94, 114, 133, 135–136
Kudreavteva, A. A., 52

**L**
Lasse, G. F., 73
Licov, A. M., 121
Liebig, J., 40, 57, 65
Livovici, M. I., 75

**M**
Marinescu, C., 66
Medvedev, V. V., 70
Mihailescu, C., xiv, 73
Mikhailova, E. A., 40, 44–46
Misustin, E. N., 66

**P**
Parton, W. I., 99–100
Pereli, T. S., 65
Pfeiffer, E. E., 119
Podari, V. D., 58–59
Pozneac, S. P., 29
Prianishnikov, D. N., 52, 57–58
Prohina, N. A., 65

**R**

Rees, R. M., 101

**S**

Scerbacov, A. P., 35, 48, 70, 91
Screabina, I. E., 20, 28, 133
Sheyin, E. V., 72
Sidorov, M. I., 7, 63
Sinkevici, Z. A., 14, 38, 53, 70
Smith, G. E., 118
Sposito, G., 15–17, 94
Stankov, N. Z., 63

**T**

Timiriazev, K. A., 74
Tulikov, A. M., 118

**U**

Ungureanu, V. G., 70–71
Ursu, A. T., xiv, 4, 7, 12, 14, 17, 19–20, 29, 38, 42–44, 46, 60, 63, 70, 73, 78–82, 86–87

**V**

VandenBygaart, A. J., 44, 128
Vassenev, I. I., 48, 70, 91
Vernadsky, V. I., 15, 27, 63, 66
Viliams, V. R., xiii, 5, 7, 16, 18, 39–40, 51, 135
von Thaer, A., 40

**Z**

Zagorcea, K. L., 43, 63
Zakharov, I. S., 66
Zveagintev, D. G., 52

# Subject Index

**A**
Actinomycetes, 63
Adsorption capacity, 33–35
Aggregate-size distribution, 135
Agro-ecology, 106
Agro-ecosystem, xiii, 17
Animal burrows, 43, 63–65
Arable crops, 5, 34, 42, 45, 61, 79, 100–102, 106, 114
Archaeology, 131–136
Available water capacity, 72–73, 83, 94, 121

**B**
Bacteria, 16, 63, 65–66, 70, 99
Balti steppe, 6–7, 60, 80, 83
Base exchange, 33–34
Bessarabia, 5–7, 43, 58, 73, 118
Biodiversity, xiii, 3, 63–66, 87, 93, 99, 114, 137
Biological cycle, 17, 99–103, 109
Biosphere, viii, xiv, 12, 15, 17, 60, 66, 107
Black Sea, 5, 81, 132
*Bonitet* rating, 19, 79, 81–82, 85, 87, 89, 136
Brown forest soil, 12, 14
Buffer capacity, 33
Bulk density, 40, 44–46, 70–71, 73–74, 102
Buried soils, 21, 133–134

**C**
Calcium, 29–30, 33, 37–38, 45–46, 48, 54, 58, 78, 84, 102–103, 119
Calcium carbonate, 78, 80, 133
Canada, 11, 44, 128
Carbon
    emissions, 42
    fixation, 17
    sink, 39–49, 52
    stocks, 128
Carbon cycle, xiv, 17, 40–41, 100
Carbon:nitrogen ratio, 42, 100–101, 118
Catch crop, 114
Catena, 11, 15–16, 69
Central Asia, xiv, 65
Chaco, 11
Chernozem evolution, 135

Chernozem soils
    Calcareous (Calcic), 77, 79–81
    Common (Haplic), 81, 83, 85
    Leached (Haplic), 85–86
    Podzolised (Luvic), 86–87
    Southern (Calcic), 66, 78
    Typical (Haplic), 83–85
    Xerophyte-wooded (Vermic), 81–83
China, 78, 94
Clay
    loams, 6, 19–20, 54, 72
    minerals, 15, 28, 33, 133, 135
Climatic change, xiii, 41–42
Climatic regulation, xiv
Colloidal fraction, 20
Compost, 4, 55, 116, 118
Conservation farming, 14
Crop rotation, xiv, 4, 6–7, 14, 49, 55, 74, 81, 100–102, 110–118, 122–123, 126–128, 137
Crop yields, 4, 43, 58, 69, 87, 105, 121–122, 124–128, 137
Crotovina, 63–64, 82, 133

**D**
Dark grey forest soil, 12–13, 48, 78
Decomposition, xiv, 17, 39–41, 52, 57, 99–101, 110, 116
    rate, 101
Desertification, 3–4, 73
Drainage, 4, 7, 15–17, 37, 60, 65, 71–72
Drought, 6–7, 69–75, 79, 110, 118, 122
Dust Bowl, 4, 94

**E**
Earthworms, 14, 64–65, 83, 107, 116
Eastern Europe, xiii, 48, 87, 92, 94, 105–106
Ecological agriculture, xiii, 4, 14, 17, 20, 42, 52, 57–58, 74, 94, 99–103, 105–107, 109–119, 121–128, 131–138
Ecological laws, xiv, 106
Ecosystem, xiii, 17, 29–30, 40, 71, 82, 99–100, 106, 109–110, 135
Elemental composition, 27–31, 133
Energy balance, 100
Energy budget, 103
Eroded soils, 4, 6, 14, 57, 94

Eutrophication, 57, 116
Exchangeable acidity, 33, 35, 86
Exchangeable ions, 33

**F**
Factors of soil formation, 15, 134
Fallow
    bare, 45–46, 100, 102, 115, 118, 122, 128
    ley grass, 18
Farming system
    efficiency, 121–128
    structural change, 102
    sustainability, xiv, 94, 105, 114, 132
Farm size, 106–107
Farmyard manure, 43, 55, 101, 114, 119, 122, 126, 136–137
Felspar, 27–28, 80–81, 84
Fertilizer
    application rates, 40, 122, 126, 128
    timing of application, 116
Fulvic acid, 45–48, 121, 133–134
Fungi, 16, 63, 65, 99
Future scenarios, 136–137

**G**
Gley soil, 12–13, 136
Globalization, 106, 137
Government policy, xiv
Greece, 5, 66, 93
Greenhouse gases, xiii, 41, 101
Green revolution, 7, 106
Grey forest soils, 12–14, 48, 53–54, 78
Gypsum, 14–15, 33, 60, 81

**H**
Herbicide, 102, 116, 119
Historical grain yields, 6, 122
Historical land use and management, 5–7
Holistic science, 119
Humic acid, 45–48, 59, 101, 133–134
Humus
    loss, 45, 52
    quality, 45–49
Hungary, xiv

**I**
Illite, 28, 33, 103
Infiltration, 17, 65, 71–72, 119
Intensive row crops, 7
Invertebrates, 14, 16, 63–65, 83, 110
Irrigation, 4, 20, 29, 38, 72, 74–75, 81, 87, 106, 117

**J**
Japan, viii, 94

**K**
Kastanozem, 11–13, 64, 78

**L**
Land degradation, xiii, 4, 14, 79, 87, 94, 137
Land reclamation, 94
Land tenure, 107, 137

Land use change, xiii, 41–42
Land use planning, 89, 110, 137
Laws of agriculture and ecology, 122
Leaching, 14–15, 18, 28–29, 33, 38, 48, 53, 55, 83–85, 101–102, 116, 133, 135
Legumes, xiv, 52, 59, 66, 70, 114, 118, 122, 126, 128
Livestock husbandry, integration with arable, 114, 137
Loess, 43–44, 80, 93, 133
Long-term field experiments, xiii, 100–102, 111, 121, 123–126
Lucerne, 17, 65, 81, 118, 128

**M**
Magnesium, 33, 38, 46, 84, 103
Mexico, 11
Mica, 27–28, 80–81
Micro-organisms, 52, 63, 65–66
Mineralisation, 17, 55
Minimum tillage, 4, 116–117, 126–128, 136
Moldavia, 5–7, 14, 19, 30–31, 35, 38, 42–43, 45, 47–49, 53–54, 58–60, 70–71, 73, 110, 132–133, 135
Moldova, xiv, 3–4, 6–7, 12, 14, 19–20, 27, 29, 37, 40, 43–45, 48, 52, 58–61, 64–65, 70, 73–74, 77–82, 86–87, 101, 107, 110, 122–123, 132, 136
Monoculture, 102, 109, 114–115, 117–119, 122, 126
Mulch, 116, 119
Mull humus, 81, 83

**N**
Nitrate
    crop- uptake model, 55, 102
    leaching, 53, 55
    sink, 53–54
Nitrification, 52, 55
Nitrogen, 6, 17–18, 40, 42–43, 45–46, 49, 51–55, 57–58, 66, 93, 100–102, 116, 118, 122, 126, 136
    fixation, 101
Non-renewable resources, xi, xiii, 105–106, 137
North Caucasus, xiv, 63
Nutrient
    availability to crops, 103
    budget, 60, 103, 118
    cycle, xiv

**O**
Organic farming, xiii, 40, 136
Organo-mineral complex, 11, 15–16

**P**
Pampas, xiv, 11, 78
Particle density, 69–71
Particle-size distribution, 11, 132
Pathogen, 99, 117–118
Pedosphere, 11, 15–16
Pesticide, 3, 15, 99, 119
pH, 33–35, 38, 77, 86
Phaeozem, 77–78
Phosphate
    biological pumping, 58
    fertilizer, 57–58, 60
    global shortage, 57

# Subject Index

Phosphorus, 17–18, 27, 29, 38, 39–40, 57–61, 90–92, 103, 122, 126
Plant nutrients, 40, 93, 99, 101, 121
Pollution, 7, 15, 18, 116, 118, 137
Polyfactorial trials, 121
Pore space, 18, 69–71
Porosity, 18, 63, 65, 69–71, 74, 81, 85
Potassium, 18, 27–28, 33, 39, 51, 58, 60, 90–92, 103, 122, 126
Prairies, xiii–xiv, 11, 77–78, 119, 122
Precision farming, 89–92
Public health, 118

## R

Reaction, 33–35, 79, 81, 84, 103, 137
Romania, xiv, 6, 58
Root activity, 20, 37–38, 48–49, 63, 83, 99, 103, 106, 110, 116–118, 122
Rural community, 107, 137
Russia, xiv, 4–6, 11–13, 39–40, 42, 44, 70, 78

## S

Saline soils, 6, 14
Salinity, 4, 14–15, 37–38, 87
Sand, 19–20
Silt, 19–21, 28–31, 70, 72, 74, 77, 85
Smectite, 28, 33
Social organization, 106–107
Societal values/perspective, xi, xvii, 40, 93–94, 110, 132, 137
Sodic soils, 14–15, 87, 94
Sodium, 14, 33–34, 38, 45, 102
Soil acidity, 33, 35
Soil aggregate, 16, 20, 71
Soil classification of Moldova, 11
Soil compaction, 87
Soil conservation, 4, 94
Soil degradation, xiii, 4, 35, 107, 122
Soil erosion, 5–6, 14, 17, 73, 87, 106, 114, 119, 136
Soil evolution, 28, 132
Soil fertility, xiii–xiv, 3, 39–41, 63, 99, 105–106, 114, 116–118, 122, 136
Soil geography, 11–15, 78–86
Soil health, 105–107, 109–119, 122, 137
Soil horizon, 15–16, 37
Soil landscape, 15–16
Soil Law, 4, 137
Soil maps, 12, 89–90
Soil micromorphology, 83–85
Soil mineralogy, 27–31
Soil morphology, 44

Soil organic carbon
  depletion, 45
  gain, 128
  labile, 45
Soil organization, 16
Soil parent material, 20, 94, 110, 133
Soil permeability, 73
Soil productivity, 4, 30
Soil profile, 14–17, 20–21, 42–44, 58, 69, 78–82, 84, 89–91, 134, 136
Soil quality, 105–107
Soil resilience, 4, 17
Soil respiration, 65–66
Soil science, viii, xiii, 5, 7, 15, 27, 94, 101
Soil solution, 37–38
Soil structure
  assessment, 19–24, 70–75
  formation by grass roots, 70
Soil texture, 19–24, 72, 87, 94, 133
Soil variability, 89–92
Solonetz soil, 4, 14, 33, 87
Soluble salts, 37–38
Soviet Union, 7, 58
State (condition) of agriculture, xiii
Steppe, xiii, 5–7, 11–13, 27, 43–45, 64–65, 72–73, 78–83, 94, 110, 132–133, 137
Sulphate, 33, 60
Sulphur, 18, 29, 57–61, 119
Sustainability, viii, xiii, 17–18, 100, 107, 109, 113, 115, 128, 137
Sustainable agriculture, 100, 119

## T

Tillage, xiv, 4, 6, 20, 106, 110–111, 116–117, 126–128, 136
Total chemical analysis, 29–31
Trajan's Bank, 81, 133–135
Turnover of soil organic matter, 40

## U

Ukraine, xiv, 6, 11–13, 37, 42, 44, 64, 70, 80
United States of America, 4, 11, 92, 94

## W

Water cycle, xiv, 17
Weathering, 27–29, 41, 60, 80, 103
Western Europe, 7, 92, 94, 136–137
World Commission on Environment and Development, 137
World population, 131, 136
World Reference Base for Soil Resources, 12, 78
World Soil Charter, 16